THE GOLDEN HARVEST

A History of Tobacco Growing in New Zealand

Patricia K. O'Shea

HAZARD PRESS
publishers

First published 1997
Copyright © 1997 Patricia K. O'Shea
Copyright © 1997 Hazard Press

The author asserts her moral rights in the work

This book is copyright. Except for the purposes of fair reviewing,
no part of this publication may be reproduced or transmitted in any form
or by any means, electronic or mechanical, including photocopying, recording,
or any information storage and retrieval system, without permission
in writing from the publisher.
Infringers of copyright render themselves liable to prosecution.

ISBN 1-877161-29-2

Published by Hazard Press
P.O. Box 2151, Christchurch, New Zealand
Production and design by Orca Publishing Services Ltd

Printed in Malaysia

CONTENTS

Author's Note	5
Acknowledgements	6
Foreword	7
Introduction	9
Chapter One: The Dreamy Weed Takes Root in New Zealand	12
Chapter Two: Pioneers and Entrepreneurs	17
Chapter Three: Motueka Farmers take a Chance	23
Chapter Four: How It Was Done	29
Chapter Five: Companies and Controversy	37
Chapter Six: Trying Times and Trips to Wellington	45
Chapter Seven: Trials and Tribulations in the Pongakawa Valley	55
Chapter Eight: A False Sense of Security?	60
Chapter Nine: 'The Fight of Their Lives'	67
Chapter Ten: Yelling for Recognition	78
Chapter Eleven: Growth, or Empire Duty?	91
Chapter Twelve: Battles on All Fronts	109
Chapter Thirteen: Changes in the Wind	126
Chapter Fourteen: Miracle or Mirage?	136
Chapter Fifteen: Gloves Off	149
Chapter Sixteen: Bill and the Bill	166
Chapter Seventeen: Savage Surgery	182
Chapter Eighteen: Death of the Dream	198
Raking Through Ashes	209
Officers of the Tobacco Growers' Federation 1938-95	213
Members of the Tobacco Board 1936-86	213
Tobacco Production in Motueka 1925-95	214
Average Price of Flue-cured Tobacco Leaf	216
Sources of Reference	217
Index	219

Author's Note

The initial intention of this book was to record the lives of those who have been involved in the tobacco-growing industry in New Zealand. At first I thought this would be quite a short, personalised story. As my research continued, however, I found it would be impossible to do justice to those who have spent their lives in the industry without recording the political and economic undercurrents which have washed over the industry in its eighty-year history.

I have, therefore, attempted to combine the personal experiences, so freely shared by those I approached, with the written and documented evidence of the industry. The result is probably not what many will expect. To those who expected a more personal account, complete with stories (true and legendary) of growers' and workers' exploits, this may be a disappointment. Despite that, I hope readers will find this to be at least the bones of the real story. If it proves to be simply the beginning of a series I will be delighted. As those who have grown tobacco will know, there are many more stories to be told. Reading this may spur someone to set them down.

Acknowledgements

The greatest thanks go to those who willingly shared their memories, photographs and family documents – without them, this book would not be possible. Every one added something to the story. Thanks to the New Zealand Tobacco Growers' Federation for providing the opportunity and the funding to begin this project. The federation features largely in this story, not always in a flattering light, and they 'took a punt' in letting me loose on their material. My thanks also to the New Zealand Tobacco Industry Historic Trust for their support as I worked. Very special appreciation to Graham Dunstall, of the University of Canterbury History Department, for his encouragement and helpful suggestions.

Sincere thanks to the Nelson Provincial Museum, the National Archives, Wellington, Rothmans, for access to their archives in Auckland, and the Motueka Museum. Special thanks to Robert Lowe, of Wellington, who so generously made his dining room available to a stranger, along with extensive material on Charles Lowe. And my grateful thanks to Zane Ostergaard, at Motueka Pharmacy, for his skill and patience in producing what seemed like an endless stream of photographs.

Thanks also to those who contributed financially, for making the project possible:
- New Zealand Tobacco Growers' Federation
- Trustbank Community Chest
- New Zealand Lotteries Grants Board
- Rothmans of Pall Mall (NZ) Ltd
- Tasman District Council
- Westpac Banking Corporation, Motueka, which supported this project in recognition of the role tobacco growing has played in the economy of the Motueka district over its seventy-year history

Finally, many thanks to my family and friends for their patience through this project. As I have lived and breathed nothing but the tobacco industry for more than two years, they have been exceptionally supportive and at least pretended to be interested even if they were bored stiff. Thanks, folks.

Foreword

Tobacco. Time was when the very word conjured up images of far-away, exotic places and an atmosphere of power redolent with the pungent and, to many, aromatic perfume of richly blended cigar or pipe smoke. In the Western world, at least, smoking was largely confined to men, with all the social separation that implied in pre-feminist days – cigars smoked after dinner, away from the delicate sensibilities of the ladies. Or perhaps a good pipe enjoyed in comfortable, companionable silence, and even special smoking rooms and smoking jackets in which to indulge in the pleasures of the habit.

In the twentieth century the image progressed to a certain air of universal sophistication. Can we imagine the 1920s without picturing the impeccably-suited young men and the short-skirted, bobbed-haired flappers with their elegantly theatrical cigarettes and cigarette-holders? Later, the image of the down-to-earth Kiwi joker, and the equally capable Kiwi jokeress, as the typical New Zealand personality embodied the idea of laconic, hard-working country dwellers with a roll-your-own hanging from the lip, and business-like city smokers with their up-market filter tips. These images were seldom questioned and generations of young people aspired to the ideals of adult nonchalance or sophistication they implied.

By the 1960s, however, concern began to surface over the effects of tobacco smoking on personal health. Links were made between smoking and various physical disorders, especially lung cancer and heart disease. Levels of smoking were not dramatically affected but public consciousness was aroused and attitudes began to change. Going fast was the perception of smoking as a sophisticated and smart pastime – many now considered the habit unsociable and personally unattractive, even repellent.

Today tobacco smoking has become the *bête noir* of the health and social scene – try appearing cool and dashing with a long cigarette-holder these days and you are simply likely to be asked to do that outside please. In New Zealand legislation has outlawed smoking in many public spaces. More recently the government has legislated against sponsorship of sporting and cultural events by tobacco companies, thus removing both a lucrative source of funding for many organisations and an avenue of advertising for the companies. Today tobacco companies may display advertising for their products only at the point of sale, and particular emphasis is placed on preventing young people from acquiring the habit – truly a huge leap from the days when smoking was almost universally accepted as a badge of adulthood and even sophistication.

These shifting images and attitudes, whether attractive or unattractive, have been largely irrelevant to the tobacco farmers of the Motueka-based New Zealand growing industry. The hundreds of families involved, many of them into their fourth generation of tobacco growing, have been too immersed in the seasonal foibles and requirements of their crop, and the commercial and political ups and downs of the industry, to be over-concerned with the self-image of the end user. In 1994 the commercial and political implications became the factors which have now closed a more than seventy-year chapter in the history of horticulture in the wider Motueka area. No longer will locals and visitors see fields of neatly lined-up plants growing quickly to maturity, nor will they be required to provide labour for planting and harvesting this unique crop. No longer will the familiar smell of warm tobacco float over the town during the winter months as the crop is bought and reconditioned in local factories.

More significantly, there will no longer be the customary contribution to the local economy provided both through the servicing of farmers' needs, and through direct spending of tobacco-related income in the retail sector. For tobacco farmers themselves, a whole way of life is about to end, for some the only life they have ever known. This volume is offered as a record of those lives – the growers, the families, and the workers, who have been directly involved. Also in the story are researchers, tobacco companies, field officers, politicians and many others. Together they make up the story of the golden crop so long a feature of the Motueka landscape.

Introduction

The history of tobacco is surprisingly indistinct. While some believe the Chinese grew, and knew the uses of, tobacco long before Columbus ventured to the Americas, the first documented evidences of tobacco usage appear in the writings of fifteenth-century travellers.[1] The tobacco plant itself originated in South America and is a member of the nightshade family which includes petunias, potatoes and tomatoes. It is an annual plant which flourishes in latitudes from sixty degrees downward and is said to thrive on almost every soil, even the poorest. Soil type, however, plays a major part in determining the class of leaf which will be produced, so much so that the same variety grown on several differing soil types will produce many variations in finished leaf. For this reason, tobacco from different regions was used for different smoking products – that produced on dark, rich soils in Java being favoured for cigar and pipe tobacco, while tobacco from the lighter sandy soils of Virginia became the basis of the modern cigarette.

In 1492 tobacco was encountered in the Americas by voyaging Europeans who took botanical specimens back to their countries.[2] This South American variety, which can reach heights of six feet in ten weeks and produce twenty to thirty large shiny leaves, is the base species of modern smoking tobaccos.[3] English and Spanish travellers found tobacco in use in the West Indies, both for personal pleasure and in ceremonial rites and religious events – Columbus first saw cigars being smoked in what is now known as Havana. Indigenous North Americans were also known to use tobacco in certain ceremonies, burning it much in the manner of incense. Different types of pipes were used for different ceremonies, the best known being the calumet (peace pipe), and tobacco was said to induce a trance-like state in which it was believed the smoker was possessed by spirits.[4]

Smoking on any significant scale in Europe was at first largely confined to Spain. Cigars were the preferred option, although some smokers copied the cigarettes they had observed in Mexico.[5] Pipe smoking came somewhat later, possibly from contact with indigenous Canadians. With the introduction of tobacco to European life, opposing views of its value also came into being, differing very little from the present-day argument. As Portuguese, English and Spanish travellers brought tobacco back from their journeys, smoking rapidly became a popular pastime although some countries introduced extreme penalties against the practice. In China and Russia, for example, lighting up a cigar or pipe could result in execution.[6]

Until the early sixteenth century European smokers obtained their tobacco through trade with indigenous Americans but about this time small amounts began to be grown in Europe itself. The price was high, often reaching £50 per pound. By the seventeenth century, tobacco was being grown commercially in America by settlers and this replaced the trade with the indigenous population. Commercial production reduced prices dramatically to around £4.10s per pound.⁷ In England, a country often recognised more for its commercial opportunism and expertise than for punishing profitable activities, Queen Elizabeth I quickly realised the revenue potential of tobacco. Her imposition of duty of a penny per pound helped to pay for the defeat of the Spanish Armada.⁸ Subsequently, duties on tobacco remained a significant element in financing wars and were seldom reduced in times of peace. Following Elizabeth, James I was a violent opponent of tobacco smoking and raised the duty from twopence per pound to 6s 10d on American leaf, although duty on Spanish and Portuguese tobacco remained at twopence.

Charles I reversed this arrangement, being more inclined to support colonial trade. This also involved actively suppressing the growing of tobacco in England, a policy which naturally received support from American growers. The suppression of growing in England included an Act in 1652 authorising any person to enter property for the purposes of grubbing up and destroying all tobacco. English growers did not give up, however, and managed to obtain permission to continue growing by agreeing to pay threepence per pound duty on their crops; in 1654 a larger crop than ever was planted.⁹ The legislation protecting the Virginian farmers was not repealed after the War of Independence, and tobacco growing remained, technically, an offence in Britain into the twentieth century.¹⁰

Until the nineteenth century, tobacco remained the centre of claim and counter-claim regarding either its curative and pleasurable properties or its harmful and corrupting influence – and no doubt successive governments continued to collect tobacco duties. The coming of the industrial revolution brought changes to the tobacco industry. Following the Crimean War, where British soldiers had observed Turkish, French and Russian soldiers smoking cigarettes, interest began to rise in the mechanisation of cigarette manufacture in order to capitalise on the popular mass market. Continuously-fed tobacco cutting machines were already in existence, a British model having been invented in 1853 by Robert Legg. At this time, cigarettes were hand-rolled and Robert Peacock Gloag, returning to England from a term as paymaster to the Turkish forces, made the first five-a-penny brand. Known as Tom Thumbs, these were long, thick cigarettes with a straw or cane mouthpiece. Later cigarettes with vulcanite, cherrywood and even glass tips were produced.¹¹

In 1881 James Bonsack, an American student, patented the first cigarette machine, using a continuous rod system. Used initially in America by W. Duke, Sons and Company of Durham, North Carolina, Bonsack's machine revolutionised the cigarette industry. With a daily output of 250,000 cigarettes, the machines slashed the manu-facturing costs from eighty cents per 1000 cigarettes to thirty cents. In 1883 W.D. & W.O. Wills of Bristol introduced the invention to England and five years later the company was using eleven of these machines.¹² Other mechanisation followed rapidly with the invention of machines for wrapping packets of tobacco (c. 1880) and constant improvements to existing machines.

Competition amongst the myriad manufacturing companies brought mergers and takeovers, ending with the establishment of the Imperial Tobacco Company (of Great Britain and Ireland) Ltd in 1901. This joint stock company, involving thirteen of the oldest tobacco manufacturers in England, was formed to combat American moves, mainly from J.B. Duke and the giant American Tobacco Company, to capture the entire U.K. market.[13] However, after a year of intense rivalry a peace treaty was signed between the American Tobacco Company and Imperial, resulting in the formation of the British American Tobacco Company (BAT) which remains a dominant player in international tobacco trade to the present time.

Tobacco leaf to this stage was mainly obtained from the United States (for British manufacturers) but protective measures in the US created price manipulation and manufacturers began to seek alternative sources of supply. For the British it was natural to explore this option through the colonial empire, where numerous countries provided the opportunity for the expansion of commercial tobacco growing. British colonies and protectorates in Africa became the prime source of raw leaf for the British manufacturers. Nyasaland (now Malawi) and Rhodesia (Zimbabwe) proved particularly suitable and, in time, produced large volumes of relatively cheap tobacco.

In British colonies in other parts of the world, tobacco growing also became part of the range of crops of the settlers. Tobacco was grown in Australia almost from its inception as a convict settlement, and in the 1860s and 1870s quantities of leaf were produced commercially in Queensland and Victoria. Canadian farmers also took up the crop and subsequently produced almost 100 per cent of their domestic requirement. In these colonies, however, the leaf was not produced for export. Subsidiaries of British and American tobacco manufacturers were set up to process the crop domestically, blending the local product with imported leaf, mainly the aromatic American varieties. Each country's experience differed but, essentially, New Zealand tobacco production fits into this pattern.

NOTES
[1] Mack, Peter H., *The Golden Weed*, Milbrook Press, Southampton, 1965, p.1.
[2] Chippendale, F. et al, *Tobacco Growing in Queensland*, Queensland Department of Agriculture and Stock, 1961, p.1.
[3] *Tobacco Australia*, Australian tobacco industry publication, c. 1970, p.1.
[4] Mack, op cit, p.2.
[5] Chippendale et al, op cit, p.1.
[6] Mack, op cit, p.2.
[7] Chippendale et al, op cit, p.1-2.
[8] Mack, op cit, p.2.
[9] Ibid.
[10] *New Zealand Free Lance*, 11.4.28, p.7.
[11] Mack, op cit, p.12-13.
[12] Ibid, p.14.
[13] Ibid, p.15.

CHAPTER ONE

The Dreamy Weed Takes Root in New Zealand

Sealers and whalers of the early nineteenth century brought both the smoking habit and tobacco seeds with them on their journeys to New Zealand; Maori were quickly introduced to both. Reportedly taking a liking to the 'dreamy weed', some Maori were eager to cultivate the crop themselves as part of their range of horticultural produce.[1] Maori growers of tobacco are first recorded at Whakarewarewa, Rotorua, in 1839 and the crop was being cultivated by Maori in both Rotorua and Taranaki in 1868.[2] Early Pakeha settlers grew a small patch of tobacco plants which they dried, cured and processed for their own use. This tobacco was often treated in sailor fashion, being soaked in rum and molasses then twisted into tight ropes and bound with flax, a process resulting in a hard plug which could be cut for use as required.[3] In the Bay of Plenty, W. Gotch began growing and manufacturing in 1865[4] and is also recorded as growing five acres on Hay's farm at Papakura prior to 1872.[5] Gotch was not alone in these ventures as settlers in the Nelson area are also recorded as growing tobacco at this time.

Sensing that the culture of tobacco would be a worthwhile source of income for Maori, Governor Grey arranged, in 1867, for the translation into Maori of an instructional manual on tobacco growing.[6] It is clear that around the world commercial tobacco culture was then largely carried out by indigenous or enslaved people who received minimal reward for their work. Grey's manual clearly refers to certain work being suitable for 'Negroes' and also quotes the conditions of 'The Cape', probably the Cape Province of South Africa, an area known to Grey from his time there as governor in the 1850s. A proponent of the use of cheap labour for tobacco production was E. Buck, who wrote to Julius Vogel in 1878 suggesting that tobacco could be produced profitably in New Zealand if Indian 'coolie' labour were imported for the purpose. Buck was willing to arrange this and also suggested that opium and coffee might be grown under similar schemes. Vogel replied that he was personally against the use of any form of cheap labour, that New Zealand had been founded on the principle of fair reward for honest labour and that such a scheme as suggested by Buck would undermine the wages of (presumably European) New Zealanders.[7]

By the 1870s tobacco was being grown in some parts of the country on a commercial basis. In Hawke's Bay H.S. Tiffen procured tobacco seed from America and distributed it to potential growers. Most were cautious in their planting but Thomas Tanner planted twenty acres of Connecticut leaf and twenty of Havana suitable for cigars. In his enthusiasm Tanner built a large and well-equipped drying shed on his property to which other growers could also send their leaf for curing. Such was the optimism over the new crop that the local member of Parliament and Provincial Superintendent, J.D. Ormond, grew four tons of tobacco leaf.[8] Companies also began to emerge to buy, process and export local tobacco. In 1879 A. Volbracht came to New Zealand from San Francisco in order to manufacture tobacco from New Zealand-grown leaf.[9] Volbracht declared the tobacco produced in the Opotiki area to be up to 25 per cent more valuable than Australian leaf and was soon in Hawke's Bay seeking locally produced leaf for his company, the New Zealand Tobacco Growing and Curing Company.[10] Although the company failed before the purchase of any Hawke's Bay leaf, tobacco growing expanded rapidly in the early 1880s, with the sixteen acres recorded in 1881 rising to fifty acres in 1884 and again to seventy-nine acres in 1885.[11] The main areas of cultivation were Hobson, with ten acres in each of these years, Manukau, with twelve acres in 1885, and Rangitikei, where fifteen acres were planted in 1884. Numbers of growers were also recorded in the Bay of Plenty and Taranaki; one acre is recorded in the Waimea County as early as 1881 and again in 1885.

Despite the failure of his venture, Volbracht remained interested in New Zealand tobacco growing and in 1885 forwarded detailed instructions on tobacco culture to the Minister of Agriculture.[12] Volbracht was not alone in recognising the possibilities for New Zealand tobacco – the Official Record of the New Zealand Industrial Exhibition 1885 in Auckland describes displays of products made from New Zealand-grown tobacco.[13] It is not clear where this tobacco was grown but the report was confident that, with experience, New Zealand would produce the necessary varieties to manufacture the best smoking tobacco. The Auckland Tobacco Company, founded in 1883 and managed by Austin Walsh, remained in business until 1893, when it claimed that the takeover of the Dominion Tobacco Company by the American Tobacco Company had resulted in price cutting to eliminate the competition.[14] Certainly, acres under cultivation plummeted from a peak of seventy-nine in 1885 to ten in 1886, then increasing to twenty-five acres before dropping steadily to sixteen in 1891. Only four acres are recorded in 1893, when a total of 2290 pounds was produced by growers in Manukau, Kawhia and Waipa.[15] From 1896 no acreage of tobacco is recorded for the colony and references to tobacco cultivation disappeared altogether in 1901.

Tariff policies were also significant in discouraging local growing. Although the tariff structure appeared to favour domestic tobacco, there was a perception that New Zealand tobacco required to be mixed with imported leaf to produce a quality product. This meant that the mixed product was subject to only sixpence per pound less duty than totally imported leaf. This was obviously insufficient to encourage the use of local leaf on any great scale. The official view that tobacco was a revenue-producing commodity was to remain in force until well into the 1930s. Indeed, it is difficult identify a time when this view was not held. Thus despite willingness and the ability to produce a relatively satisfactory product, the infant tobacco-growing industry in New Zealand had wilted and died by the turn of the century.

Sporadic government interest in tobacco growing continued, however, and is evident from a report into tobacco growing commissioned by the Minister of Agriculture in 1901.

This study, conducted by G.F. Sutherland, concluded that certain varieties grew well in New Zealand but that the tobacco produced would need to be mixed with imported leaf to produce smokable tobacco. Sutherland described tobacco growing as 'an article of *petit* culture'[16] and thought that about one to three acres of tobacco would fit well into the programme of farms involved in small cropping. Sutherland stated that prices for New Zealand leaf ranged between sixpence and one shilling per pound, but that the unworkable excise and customs duties prevented the paying of higher prices. He suggested that an increase of duty on imported tobacco would do much to encourage local growers. South German and Austrian settlers in the Puhoi and Awanui areas were said to be willing to try out the crop but there is no evidence to show that they proceeded with this idea.

The industry was to receive its next burst of support for the local growing of tobacco with the arrival from Germany of Gerhard Husheer in 1911. Born in Bremen in 1864, Husheer had extensive experience in the tobacco industry, being articled to Messrs Ankerschmidt and Company, tobacco leaf and cigar importers and merchants, on leaving school. At twenty-five he joined Hartlaub and Company of Rotterdam and travelled widely in Europe and, later, around the world, including a trip to New Zealand during the 1890s. It was during this early visit that Husheer was impressed with the possibilities for growing tobacco here. On his arrival with his family, Gerhard Husheer set about deciding on where he would base his tobacco operations – Auckland, Hawke's Bay and Nelson were among the areas considered.

Opting for Hawke's Bay, Husheer was soon experimenting, growing his own crop at Pakipaki in 1911 and at the same time conducting field trials at the Department of Agriculture farm at Arataki, near Havelock North.[17] Early frosts and high winds proved to be a problem, as did the lack of expertise of those working with him. Unable to attend to the whole operation himself, he claimed later that the manager at the Arataki farm had discarded the infant tobacco plants and planted out nightshade seedlings instead.[18] Despite setbacks with these first attempts, there were positive reports on the future of tobacco in New Zealand. Publications such as the *New Zealand Farmer Stock and Station Journal* printed lengthy and glowing articles on the cultivation and drying of tobacco, using Gerhard Husheer's plantations in Hawke's Bay to illustrate this. Similar articles appeared at regular intervals over succeeding years, all confident that the era of tobacco production in New Zealand had arrived.[19]

Moving his growing operations to Clive Grange, Haumoana, near Napier, where he leased 120 acres, Husheer grew fifty acres of tobacco, describing it as a 'great crop, six foot high after topping. The leaves were eighteen inches long with fine, silky texture and fine bright colours'.[20] Air-curing barns, heating sheds and storage facilities were erected, and the first tobacco transplanting machines were imported from Canada. With personal finance running low, in 1912 Gerhard Husheer formed the New Zealand Tobacco Company along with local businessmen George Ebbet, Dennis Donaven, Frederick Hartshorn, William Jones, Richard Sorenson, James Cowan and landowner Waikare Karaitiana.[21] For four years Husheer harvested an average of 75,000 pounds from 100 acres, storing the air-cured leaf in bulk until 1914 when he commenced manufacturing tobacco products in a small brick factory at Port Ahuriri, Napier.

Husheer was proud of this small factory, saying he had built the 80x40ft building himself.[22] The factory was equipped with moderate, second-hand machinery purchased from business associates in Hamilton who had produced tobacco made from imported leaf. Processing and manufacturing his own leaf, Husheer considered that he produced a smokable article which he set about promoting to the public. War-time shortages meant that the product could not be packaged in tins so the tobacco was marketed in linen pouches and was known as 'Gold Pouch'. The company sent out travellers to establish a market for tobacco with a high New Zealand content – a process described by Husheer as very difficult. Even so, by 1917 the company's trade had grown to 100,000 pounds of manufactured products per year and was increasingly popular with the smoking public. 'Gold Pouch' was joined by 'Three Diamonds' and 'Red Shag' for roll-your-own smokers, and 'My Favourite' and 'Smoke-ho' flake-cut tobaccos for pipe smokers. According to Gerhard Husheer, this success brought attention from W.D. & W.O. Wills which, through its New Zealand agent, J.W. Brodie, made an attractive offer to buy out the New Zealand Tobacco Company in 1916. Husheer considered that the parent company, British American Tobacco (BAT), did not have the intention of furthering the development of tobacco growing in New Zealand and he did not encourage the negotiations.

By now the company's growing operation had expanded and 120 acres were under cultivation. Returns on investment were healthy and the company was paying out handsome dividends. In 1917, however, the company's farm at Clive Grange was struck by an attack of blue mould and growing operations were ceased, the farm closed and all the property sold at auction. The company was holding sufficient leaf stock to continue manufacturing for two to three years but needed to seek new sources of supply to maintain its business and retain the market share it had achieved. In 1918 the company was beset by problems and friction amongst its directors, adding further stress to the situation. Husheer believed he suffered from belated anti-German feeling engendered by the war – it was later said that he and his family were forcibly evicted from the factory premises by hired thugs. Subsequently the remaining directors engaged a new manager, S. Smallover, and attempted to carry on the business, although many of the staff had left following the departure of the Husheers.

Determined to continue in the tobacco business, Gerhard Husheer moved to Riverhead in Auckland, where he formed a syndicate to finance a plantation. This syndicate was backed by Auckland businessmen Enos Bond, founder of Bond & Bond Ltd, and John Burns of the Auckland firm of that name. At first, twenty acres of tobacco were planted and the resulting crop exceeded all expectation. By the time the liquidators of the New Zealand Tobacco Company approached Husheer in 1922 requesting him to resume control of that company's assets in Napier, his Auckland plantation had grown to 200 acres. As well, the syndicate was accepting leaf from private growers at Riverhead and Te Atatu.

Returning to Napier, Gerhard Husheer found there was still a large quantity of his Clive Grange tobacco left in store. He believed that, under the management of S. Smallover and a company director, Gilmour, the company had failed to produce a palatable product. Husheer took over the operation for a 'naturally low figure' and the family took some satisfaction in rebuilding the business at its original location. The factory continued to be supplied from the Riverhead plantation for some time. Not long after the move back to Napier, however, members of the syndicate began to express impatience for signs of profit and appeared unwilling to advance further money for development. Husheer then formed the National Tobacco Company, which bought out the syndicate and within twelve months the new venture was showing healthy profits. The tobacco entrepreneur took great pleasure in pointing out his success. For some years he continued to send copies of his company's balance sheet to the members of the original syndicate to underline their 'chicken-heartedness' in pulling out prematurely.[23]

By the early 1920s the company was producing several popular brands of tobacco made with a large proportion of New Zealand-grown tobacco – 'Riverhead Gold', 'Cavendish Mixture', 'Cut Plug No 10' and 'Navy Cut No 3' were all marketed in round, two-ounce tins retailing for 1s 5d to 1s 6d per tin. Gerhard Husheer and the National Tobacco Company were vital factors in promoting commercial tobacco growing in New Zealand, not only at early locations in Hawke's Bay and Auckland, but in the region which was to become the major, and later the sole, commercial producer of tobacco in New Zealand: Motueka.

NOTES

[1] *New Zealand Free Lance*, 11.4.28, p.7.
[2] Tobacco growing by Maori was observed near Rotorua by Bidwell in 1839 (J.R.Waugh, p.33). Stevan Eldred-Grigg (*Pleasures of the Flesh*, p.102) has a photograph of tobacco growing in Maori gardens at Ohinemutu in 1868, and James Belich mentions tobacco growing in Taranaki in 1868 in *I Shall Not Die*, p.33. See bibliography for details of these sources.
[3] Notes on tobacco history held by Rothmans (NZ) Ltd, Auckland.
[4] *Appendices to the Journals of the House of Representatives* (AJHR) 1880 H22 p.45.
[5] From an article on the history of Riverhead, provided by J. Husheer.
[6] White, J., *The Culture of the Tobacco Plant*, Auckland, 1867.
[7] AJHR 1878 H9
[8] Dunlop, B. and Mooney, K., *Profile of a Province – Hawke's Bay*, Hodder and Stoughton, 1986, pp 57-58.
[9] AJHR 1880 H22 p.46.
[10] Dunlop and Mooney, op cit, p.57.
[11] New Zealand Statistics 1884-85.
[12] AJHR 1885 H15A p.10.
[13] Newport J.N.W., *Footprints Too*, Express Printing Works, Blenheim. 1978, p.183.
[14] AJHR 1904 1-10B
[15] *New Zealand Yearbooks* 1885-93.
[16] Sutherland G.F., *Tobacco Growing in New Zealand*, Government Print, 1901.
[17] National Archives file LE 1 1930/16 – History of the Industry.
[18] Minutes of Evidence, Committee of Inquiry into Tobacco Growing 1930 p.526, National Archives File No LE 1 1930/16.
[19] *New Zealand Farmer Stock and Station Journal* 1913, 1914 and 1921.
[20] Minutes of Evidence, p.496.
[21] Company prospectus, courtesy of John Husheer, Napier.
[22] Minutes of Evidence, p.497,
[23] All information on Husheer's involvement with the NZTC and the formation of the National Tobacco Company taken from T. Husheer, 'Notes on the History of the Tobacco Industry'.

CHAPTER TWO

Pioneers and Entrepreneurs

Tobacco growing came early to Nelson. In 1843, on the first anniversary of the founding of the European settlement at Nelson, an agricultural show was held at which Mr McGee's tobacco plants earned special praise in the tobacco competition. In 1850 Pastor Heine planted fifty plants at Upper Moutere and in the 1880s William Brereton, of Pokororo, like many settlers around the country, was growing tobacco for use in sheep dip and for his own personal use.[1] Given the predominance of small holdings and small cropping in the river valleys in the Motueka area, it is likely that many plots of tobacco were grown for personal use throughout this time. Apart from an anonymous acre recorded in the early 1880s, the first known commercial plantings of tobacco in the Motueka area were on Horatio Everett's property at Brooklyn, on land which now forms part of the Hortresearch farm. His crops of 1888, 1889 and 1890 were sold to an Auckland company (probably the Auckland Tobacco Company), which exported them to Australia. Everett's tobacco growing then ceased. Presumably the returns did not justify continuing what was probably a risky venture. With the problems of overproduction in Australia, it is unlikely that there was any demand for small amounts from New Zealand. Horatio Everett and his family were to become tobacco growers again in the 1930s. Following this initial foray, no further activity in tobacco growing in the Nelson region is recorded until 1916, when Charles Lowe, an entrepreneurial horticulturist, decided to give tobacco a trial.

In 1905 Charles Lowe lived in Little Sidney Valley, growing and processing raspberries for export. Taking a break in a scheduled trip to London to promote his company's raspberry crop, he attended the St Louis Exhibition to see if any fruit preserving equipment was on show. However, he was fascinated by the demonstration of yellow tobacco leaf being cut up and machined into cigarettes, with packs of five being given free to onlookers. This particularly interested him as he had previously grown a little tobacco in Waikanae from seed imported from the US by T.W. Kirk, a former Director of Horticulture with the New Zealand Department of Agriculture.[2] Until his visit to St Louis, Lowe had seen only the deep brown tobacco produced through air drying and he was intrigued by the pictures of plantations and kilns which produced a golden crop. Having been delayed by his trip to St Louis, Charles Lowe decided to visit the tobacco-growing areas of Virginia and South Carolina to observe the activities himself.

In 1911 Lowe took up land in Flaxmore Road, Harakeke, where he planted out 183 acres in fruit and subdivided the area into workable pip-fruit orchards. Lowe's land in Flaxmore Road reminded him of the tobacco-growing land he had seen in the US and in 1916 he decided to experiment with tobacco. For this trial Lowe obtained tobacco seed from Gerhard Husheer at Napier. He records producing an air-dried crop of three acres, harvested in the summer of 1917. However, the crop was not purchased by Husheer's New Zealand Tobacco Company, being sold instead to J.W. Brodie, at that time agent for W.D. & W.O. Wills of Wellington. At the time, Lowe seems to have believed that Husheer could not, or would not, buy his crop.[3] Subsequently, Lowe and Husheer were never to be particularly close and some acrimonious exchanges occurred later, although Lowe was employed by Husheer at the National Tobacco Company reconditioning factory in Motueka in the early 1940s.

Lowe had proven the possibility of growing tobacco at Harakeke with his first air-dried crop. Recollecting his admiration of the yellow tobacco produced from flue curing in America, he was now keen to try out the process himself. In 1918 he arranged for Hamilton Delany, of Motueka, to erect the first flue-curing barn in New Zealand, with flues cast from iron dredge pipes.[4] This first flue barn was built into an existing fruit shed and Lowe planted an area of tobacco which was harvested and cured in the new kiln, almost certainly the first flue-cured tobacco produced in New Zealand. The quality of the crop was disappointing, however; Lowe thought his curing methods were faulty. The following year he enlisted the help of an Australian curing expert, G. Atkinson, and together they 'cured the best and largest crop of tobacco'[5] Lowe had ever grown, the product of eight acres of planting.

With seed acquired from Washington by his old friend, T.W. Kirk, of the Department of Agriculture, Lowe grew several tobacco varieties including Oronoko, White Burley and Yellow Pryor. In a letter thanking Kirk for the seed, he also requested financial assistance from the government, claiming that he had mortgaged his property to build the prototype kiln. Members of a Select Committee on Industries, set up in 1919, visited Lowe at his Harakeke property on 8 March to inspect his tobacco crop and buildings. In July Lowe wrote to the chairman of this committee, A.C.M. Wilkinson, requesting £250 towards the costs he had incurred. In addition to his kiln and machinery, Charles Lowe claimed that his expertise was now being used by others and 'surely I should be rewarded'.[6] Indeed local farmers and orchardists *were* requesting advice from Lowe. J. Addison of Ruby Bay had written to Lowe in June 1919 asking for information as he wanted to grow tobacco while waiting for his apples to begin producing.[7] Later that year Addison was included in a list of New Zealand Tobacco Company growers for the coming season, along with A. Forsyth and P. Watt of Mapua. Addison's crop was predicted to be ten acres, Forsyth's seven acres and Watt's six acres.[8] However, it seems that the company was unable to complete its contracts with these growers and they were advised to approach Lowe to make suitable arrangements for their crops, probably through his connections with Wills.

Impressed by their meeting with Lowe, and by the inspection of his crop, the select committee recommended to the Minister of Agriculture that Lowe be appointed as an expert to advise prospective tobacco growers.[9] Nothing came of the Minister's promise to give full consideration to this recommendation, or to Lowe's request for financial assistance. Undeterred, Charles Lowe continued with his enterprise. Six barns of his 1919-20 crop had been cured and stored in the shed by 24 April 1920. The curing of the seventh barn was completed on that day and the furnace fires damped down. But despite precautions and several checks, the barn and the shed were destroyed by fire during the night of the 24-25 April. The whole crop was lost as well as quantities of Delicious apples and various items of machinery, some delivered only on the previous Friday and never used. Lowe was convinced that this fire was the result of 'a deliberate act of incendiarism',[10] but it is not clear who he thought may have been responsible for such an action.

W.D. & W.O. Wills financed replacement kilns on Lowe's property, across the road from the burnt-out shed, advancing £400 towards the cost. Two kilns were built in 1921 – 20ft-square buildings heated by steam boiler. Eighty steam pipes, each eighteen feet long, with vents, were used to cure the crop. The system was easy to control, but Lowe described it as expensive, saying that the pipes in each barn cost more than a whole kiln in 1938.[11] During 1920 Lowe, in conjunction with T.J. Whittaker of Wills, contracted at least two growers to grow tobacco on an experimental basis at a price of twopence per plant for best-quality tobacco leaf. It was proposed to cure this leaf in Lowe's kilns at Harakeke and Wills would presumably buy the finished product.

W.D. & W.O. Wills (NZ) Ltd had begun manufacturing tobacco products in New Zealand in 1919 at its factory in Wellington. The company was described in 1930 as a private company registered at Wellington with a nominal and subscribed capital of £830,000. The British American Tobacco Company Ltd (BAT) of London held significant shares in this company, as it did in the Imperial Tobacco Company Ltd of NZ. Wills itself held the bulk of the shares in the Dominion Tobacco Company Ltd, which BAT had taken over in 1894. Until 1920 Wills had imported practically its entire tobacco requirement as its market was concentrated on cigarettes, for which it needed the brighter tobacco produced by flue curing. Lowe's experiments were therefore of some interest to Wills, although the company's overseas connections appeared to favour the use of tobacco from established American and British colonial producers. Lowe's relationship with Wills was, therefore, ambivalent. Having sold his early crop to the company, he was nevertheless sceptical of its overall attitude. Lowe felt that, despite its support of his endeavours, Wills was generally inclined to discourage tobacco growing in New Zealand. Subsequently Wills withdrew its support for Lowe. He continued growing tobacco at Harakeke until 1923 when, following his financial problems, the company demolished the flue barns and removed the equipment to Wellington.[12]

International trade factors were also affecting the tobacco industry in New Zealand. In January 1920 prices of tobacco products were said to be at levels amounting to profiteering, having increased by a third since the end of the First World War. In February the Australian tobacco crop suffered a severe failure, and in March imported tobacco from the US was quoted as fetching six shillings per pound landed and without the addition of customs duties.[13] These developments probably stimulated tobacco companies to encourage the growing of leaf in New Zealand. Local growers were receiving a maximum of two shillings per pound before excise duty – an attractive proposition to those companies using New Zealand leaf for the manufacture of cut tobacco, as they were required to pay only excise duty on releasing the tobacco from bond. The expansion of production in Rhodesia was not yet under way and, in addition, companies were said to be committed to encouraging the growing of tobacco in those countries where they had established manufacturing facilities.[14] This had seen tobacco growing fostered in Australia, and certainly Wills had purchased the occasional crop from local growers through its New Zealand agent, the British Empire Trading Company. Conditions were beginning to favour another speculative sortie into commercial tobacco growing in New Zealand.

By 1920, when Charles Lowe's first kiln was burnt down, other local farmers were becoming interested in the potential of tobacco. Knowing little or nothing of Lowe's activities, Cecil C. Nash of Brightwater was introduced to the idea of tobacco growing by F. Hill of E. Buxton and Company of Nelson. Along with Charles Lowe and Gerhard Husheer, Cecil C. Nash was to be a key personality in the promotion of tobacco growing in the Nelson-Motueka region. Born in Nelson in 1874,[15] Nash eventually purchased what was known as the Rutherford Farm at Brightwater, the site of his early tobacco trials. Nash received instruction in the sowing of tobacco seed in beds from Australian tobacco expert and NZTC director Gilmour, who had previously worked for Wills for thirty-two years. Although Nash's first attempt failed, his second sowing was very successful and he then planted half an acre of air-dried tobacco in 1920, which was to be sold to the NZTC. This company had earlier approached Buxtons with proposals to encourage tobacco growing in the Nelson area but the idea did not attract Buxtons at that time.[16] The company's representative, Fred O. Hamilton, had put Gilmour in touch with farmers who might be interested in the idea – Cecil Nash was one of these and was the only one to pursue the project to completion.

Nash saw the potential of tobacco and began promoting the crop immediately. Local agricultural shows provided an opportunity to display plants and leaf to the general public and Nash took advantage of them as early as the summer of 1920-21, when he had a stall at the Brightwater Show. However, the NZTC went into liquidation whilst Nash was producing his first tobacco crop, leaving him with his leaf on his hands. Fortuitously, he received a letter from Gerhard Husheer expressing interest in his tobacco. He forwarded the leaf to Husheer, who had returned to Napier having taken over his old company's assets, and he was delighted to receive £50 for it, this being sixpence per pound more than his original contract with the NZTC.[17]

Nash's success with his first tobacco crop, and his undoubted enthusiasm and belief in the future of tobacco, prompted him to attempt to set up his own manufacturing business in Nelson.[18] This attempt apparently came to nothing. However, Husheer intimated that he would take all the tobacco Nash could supply and suggested that Nash try for eighteen acres in the coming season. In the face of farmer scepticism, Nash succeeded in procuring only eight acres, six of which he grew himself, the other grower being Griffith Lewis of Hope.[19] Recognising the value of Nash's enthusiasm for tobacco growing, Gerhard Husheer appointed him as the National Tobacco Company representative and instructor for the district. In this role Nash travelled the rural roads, on a bicycle at first, holding meetings and talking tobacco to any farmer who would listen.

Nothing discouraged Cecil Nash. Many recall him cycling around the district in the early days, when company funds were limited, as he gradually convinced growers in the many valleys of the county to plant a little of the exotic crop he was determined to foster. By sheer enthusiasm and strength of personality, he won over the hard-working and sceptical small holders who had been desperately surviving in the post-war era of erratic and unstable prices for their goods. Nash's own sense of loyalty demanded a similar commitment from 'his' growers: his comments and dealings with farmers suggest that 'if you weren't for him, you were agin him'. His dedication to the industry, however, could never be questioned. Company advisers developed unique relationships with their growers. One grower recalls Cecil Nash as 'a bombastic old coot' – others acknowledge that, while what he said was law, he knew what he was talking about. When his bicycle was replaced with a car, Nash attached lists of instructions to the windows for growers' information, including the order to 'keep them turkeys out of the tobacco'.[20] Whatever his personal idiosyncrasies, Nash was always available to growers and responsive to their requests for help.[21]

Cecil Nash came to be seen as the leading adviser in the district and, while he always deferred to Gerhard Husheer, he was often critical of other 'experts'. Nash tells of visiting Charles Lowe's drying kilns at Harakeke and, while he was impressed with the facilities, he was totally unimpressed with Lowe himself. Nash found a heap of kiln-dried tobacco rotting behind the sheds and 'fished some out from the middle',[22] finding it equal in quality to any similar leaf he had seen. This experience, typical of several related by Nash, together with his fierce loyalty to Gerhard Husheer, forged an acrimonious relationship between Lowe and Nash. This was reinforced by Lowe's later association with Wills growers and the divisions amongst growers, which appeared to fall in line with company affiliations.

Cecil Nash was almost fanatical in defence of tobacco growing. In later years he maintained that he had been approached in December 1924 and offered a great deal of money to discourage tobacco growing in the area. Nash claimed Husheer put a proposal to him, while another man hovered in the background, which involved payment of £35,000 for Nash and £65,000 for Husheer. Nash refused the offer point blank, asserting that he gained more pleasure from travelling around the countryside 'seeing happy and contented growers' than he would from having his 'arms full of money'. In 1935, on a visit to Husheer's home in Napier, Nash claimed the matter was raised by Husheer who was upset when Nash said he had no regrets that he had turned down such a huge sum of money.[23] The lamppost outside the Masonic Hotel in Nelson, which Nash is said to have leaned against while this incident took place now watches over his grave at Nelson's Wakapuaka Cemetery.

Today there is still local speculation on who was behind this alleged event – guesses range from a rival tobacco company to someone who wished to see tea growing promoted instead. The incident as reported does not sit well with Husheer's consistent record of support for growing tobacco in New Zealand and his rejection of previous attempts to buy out his company. Whoever was behind it, the attempt to stifle the infant tobacco-growing industry failed and both Cecil Nash and Gerhard Husheer remained involved in the tobacco industry for the rest of their working lives.

NOTES

1. Beatson, K. and Whelan, H., *The River Flows On*, Beatson and Whelan, Nelson, 1993, p.51.
2. *Motueka Star-Times*, 28.7.38.
3. Lowe maintained later that Husheer filed for bankruptcy in 1917 but he may have misinterpreted the situation. At this time Husheer was being manipulated out of his company and Lowe produced insufficient tobacco for the new directors to be greatly interested. From personal papers of Charles Lowe.
4. Hamilton Delany built many more kilns late in the 1920s. Numbers of these can still be seen in the area. Some were converted to the down-draught system of curing tobacco in the 1950s and 1960s.
5. *Nelson Evening Mail*, 23.11.21,
6. Charles Lowe's submission to Select Committee on Industries, 1919. National Archives, Wellington.
7. Select Committee on Industries, 1919.
8. Ibid, list of NZTC growers, 1.9.19. The company's own plantation was to be sixty acres, the Rushton brothers would produce seven acres and Taylor of Tongariro (listed as a Native [Maori]) was to grow five acres. With total production estimated to be 85,000 pounds, the company obviously intended to continue operating, although a request for official assistance through a rebate of duty on the 50-60,000 pounds it had on hand indicates it may have been struggling at this point.
9. Ibid.
10. Report of Charles Lowe's bankruptcy case, *Nelson Evening Mail*, 23.11.21.
11. *Motueka Star-Times*, 28.7.38. Cecil Nash claimed in 1930 that the new kilns had cost £1000.
12. Op cit.
13. *Nelson Evening Mail*, 17.2.20, p.7.
14. Comments by former Wills manager, Motueka, A.L. Black, on video compiled by Execam, Wellington.
15. Cecil Nash was the son of a Nelson policeman, John Nash, who was allocated the number one when the New Zealand Police adopted a numbering system in the 1870s.
16. Minutes of Evidence, Committee of Inquiry into Tobacco Growing, 1930, p.632.
17. Brereton, C.B., *Vanguard of the South*, A.H. & A.W. Reed, Wellington, 1952, p.166.
18. Minutes of Evidence, Committee of Inquiry into Tobacco Growing, 1930, p.152.
19. A contemporary report says that thirty acres were sought but only twenty grown. *Nelson Evening Mail*, 21.4.22. Interestingly, none of this activity was recorded statistically; in fact according to New Zealand Statistics, no tobacco was grown in New Zealand at all in the years 1919 and 1921-22. As Gerhard Husheer's syndicate was still operating in Riverhead during this time the records are misleading.
20. Interview with former grower A.P. (Pat) Beatson, Orinoco, 3.8.95.
21. Minutes of Evidence, Committee of Inquiry into Tobacco Growing, 1930, p.224.
22. Ibid, p.4. Nash dates this visit in 1924 but as Wills removed the kilns in 1923 – probably after the 1922-23 curing season this date is suspect.
23. Details of this story from a letter written by Nash to Lance Mytton, 28.6.51. This episode is now somewhat legendary and the facts are difficult to verify.

CHAPTER THREE

Motueka Farmers take a Chance

In 1922, at Nash's instigation, several farmers in Riwaka and the Motueka Valley tried out the new crop which could offer cash returns. A group of Riwaka farmers grew eighteen acres of tobacco for the National Tobacco Company and received an average of two shillings per pound. This group, Joshua Parkes, Percy Bate, Ralph Bate, Billy Owen and Fred Cook, were all well known in the area. Joshua Parkes was later to be a tobacco adviser for the National Tobacco Company[1] and in 1926 Billy Owen was to convert his hop kiln to become one of the first producers of kiln-dried (flue-cured) tobacco.[2] Possibly also in 1922-23, Frank Moss grew about two acres in the Little Sidney Valley, also for the National Tobacco Company. This air-dried crop was hung to dry in the loft of the hop kiln, the plants being turned regularly for even drying and the leaves shaken to discourage the growth of mould.[3]

At Tapawera F.W. Gibbs commenced growing around the same time, having a contract for two acres of air-dried tobacco with the National Tobacco Company. In 1923 Lance Mytton began growing about three acres of tobacco on his brother's land in the Motueka Valley just south of the Pearse River.[4] In the same year, the Barker family planted their first crop at Stanley Brook, and in 1924-25 E.C. (Elliot) Fry is recorded as a grower in Riwaka. During this early period Manoy and Sons of Motueka was the agent for National Tobacco, making advances on crops to growers on the company's behalf and attending to the freighting of harvested tobacco to the factory at Port Ahuriri,[5] where assessment of quality and price were made. The lack of local facilities for this purpose was to be of major concern to growers until well into the 1930s.

By 1925 tobacco growing was beginning to take hold in the Motueka Valley, especially around Ngatimoti and the Orinoco Valley. Cyprian Brereton had previously grown a little tobacco with seed he brought back from the First World War, and was definitely aware of the possibilities of the crop for the area. Prior to the popularity of motor vehicles, Orinoco farmers had received a significant part of their income from growing chaff for horses[6] but now the need for a replacement crop was becoming urgent for their survival. At meetings around the area Cecil Nash promoted tobacco growing. Many were doubtful of the promised returns, which were estimated at £50 per acre, privately thinking that Nash 'had a bee in his bonnet and ought to be in the mental'.[7] Even so, all eleven farmers present at the Orinoco meeting signed contracts with Nash, as did others in surrounding areas.

By 1925-26 the area under contract exceeded 150 acres. The formerly sceptical farmers were now fully aware of the potential of tobacco growing and keen to participate in the industry. Although it had ceased growing on its plantations at Riverhead, National Tobacco was unable to offer contracts to all those wishing to grow tobacco. To some the only other option appeared to be the fostering of an export trade in New Zealand-grown tobacco. Led by Charles Lowe, a group of would-be growers, mainly from the Motueka-Riwaka area, called a meeting to which they invited the Director of Horticulture from the Department of Agriculture, J.A. Campbell. The meeting was held in the Anglican Church schoolroom in Motueka on 27 June 1926 and an association of growers was formed under the chairmanship of Charles Lowe. The government, keen to support the export initiative, offered support by way of guaranteed prices on the following terms:-

> kiln dried tobacco was to be lemon coloured;
> other tobacco was to be not darker than mahogany;
> the guarantee was to be for three years;
> growers must comply with instructions;
> prices were to be 2s 3d lb for flue-cured, 1s 4d lb for air-dried.[8]

Cecil Nash attended this meeting as a representative of the National Tobacco Company and was less than enthusiastic about the idea. Nash asserted that there was no tobacco expert in New Zealand other than Gerhard Husheer, and that New Zealand could not produce exportable leaf. He warned that any National Tobacco Company growers who took part in the export scheme would lose their contracts.[9] He believed that if growers attempted to grow tobacco for both the local and the export markets they would naturally submit their best tobacco for export, leaving only second-grade tobacco for domestic manufacturers.[10] 'No man can serve two masters' was Nash's customary response to those National Tobacco Company growers who expressed interest in the export proposal.

Despite Nash's opposition and the implied threats, the association proceeded and 1926 was a busy year for Lowe and the thirty members. A committee of growers was elected and a receiving depot was established in High Street, Motueka, in a building purchased from Sidney Rowling of Riwaka.[11] Thirty-eight growers were contracted under the government guarantee system and several hop kilns were converted for the purpose of flue curing. Eight to ten tons of tobacco were produced by these growers but the leaf was never exported. Between 1920 and 1926 Buxtons of Nelson had approached W.D. & W.O. Wills many times about purchasing Nelson-grown leaf, but the manufacturing company was not interested until the possibility of significant amounts of flue-cured leaf became a reality.

After the association's crop was harvested in 1927, Fred O. Hamilton of Buxtons forwarded samples to Wills. Wills appeared impressed with the leaf and sent samples to Australia for testing. Very quickly a cable was received instructing Wills (NZ) to buy the whole crop if it were available. Wills promptly made an offer to the association to purchase all suitable leaf from this crop at the equivalent of the government guaranteed price.[12] The Department of Agriculture was against this proposal as the crux of its project was the fostering of growing for export. J.A. Campbell, Director of Horticulture for the department, was already expressing the opinion that there would always be trouble with only two buyers of domestic tobacco leaf.[13] However, Wills' offer was made to the association and not to the department, and Wills' nett price in New Zealand equalled the gross price expected for the shipment in London. In addition Wills promised to contract to purchase the whole of the following season's crop, which was expected to be approximately seventy tons.

Wills also agreed to send a sample shipment of leaf to London for evaluation. Seven hundred pounds of the association's first crop was dispatched and, while of good quality, was said to have been poorly handled and packaged. The following year a larger shipment of 5000 pounds of flue-cured tobacco and 5000 pounds of air-dried was forwarded. This received very positive reports, with bright flue-cured leaf assessed as being the best colonial tobacco tested and close to medium Virginia leaf. Some technical and production faults were present but it was predicted that these would be overcome with experience. The air-dried leaf forwarded also received a good report, although its colour was said to be rather reddish.[14] It is interesting to note that W.D. & W.O. Wills prepared and packed the export samples in New Zealand and it was this firm's London connections who provided the reports on their quality at the receiving end.[15]

The National Tobacco Company had concentrated on purchasing air-dried tobacco and was understandably disturbed at competition from government guaranteed prices for kiln-dried leaf for export. As well, Nash was indignant that the association tobacco had not been offered on the open market – he questioned why Wills had been given preferential treatment.[16] Even so, Nash did purchase tobacco from a number of association growers whose tobacco was rejected by Wills. One grower was said to be bedding down pigs on his rejected tobacco and Nash purchased his surplus leaf for 1s 6d per pound. In all, seventeen association growers sold tobacco to National Tobacco following the 1926-27 season. Nash remained convinced that Wills was attempting to destroy the National Tobacco Company and that Lowe was acting in the interests of Wills.[17]

The essential result of the 1926-27 season was the existence of two purchasing companies for New Zealand-grown tobacco. Following this season Wills offered growers contracts for three years for kiln-dried tobacco and one year for air-dried tobacco. Numbers of growers leapt immediately from 140 in the Motueka-Nelson area in 1926-27, to 314 for the following season, rising again to 559 in the 1928-29 season. With air-dried tobacco already widespread in the Motueka Valley, Orinoco and Pokororo, the introduction of flue curing saw Riwaka acreages expand rapidly, while Dovedale and other outlying valleys adopted the new crop on a large scale. Kilns could be financed through Wills, with the advance being repaid from the proceeds of the crop over two or three years. Some National Tobacco growers changed to growing flue-cured tobacco for Wills, attracted by the contracts and the financial offer – one of these, H.M. Helm, built what was probably the first kiln in Pangatotara.[18] In the 1927-28 season Wills contracted for an area of 200 acres, and financial assistance for the erection of kilns and bulk sheds was provided to approximately twenty growers.

By the end of this season between twenty-five and thirty kilns were being used from Riwaka to Wai-iti.[19] In the following season 400 acres were contracted for, with kiln numbers rising to sixty-nine.

Like the National Tobacco Company, Wills provided instructors to assist growers with the culture and curing of flue-dried tobacco. In fact, during these first three years of its operation in the Motueka district Wills provided different instructors each year, a matter which became of great concern to the growers. Each instructor, it was said, brought different ideas on the correct way to manage the crop. In the first year, 1927-28, instructors Whittaker and Sedrick Brame, both fresh from the company's China station, told growers to remove the suckers, or laterals, in order to encourage growth into the main leaves. In 1928-29 Lough, from Australia, instructed growers not to 'lateral' at all in order to produce a finer, lighter leaf. The following year, 1929-30, another Australian expert, Gilmour, proved just as adamant on the issue of not lateralling. Sedrick Brame, who had wide experience as a tobacco adviser in the US, Canada, Russia and Japan, was exceedingly critical of this 'no-lateralling' advice. He believed that methods used in Australia to combat a particular soil taint were being forced on New Zealand growers who had no such soil taint.[20] By 1930 growers were thoroughly confused and requested that a permanent government instructor be resident in the area to provide consistent advice.

This was not the first time Wills' experts had been involved in giving controversial advice to Motueka growers. In 1926 Brodie and Lough (then London manager of Wills) had visited the farm of Cyprian Brereton in Orinoco, making disparaging comments about the climate and the crop to the farmer. Commenting that the climate was impossible owing to too much hail and wind in summer, as well as frosts in spring and autumn, they told Brereton 'in the most impressive way possible, "for his own benefit" '[21] that his crop was worthless. Cyprian Brereton received £253 from the National Tobacco Company for this same tobacco. He also claimed that the company experts had quietly agreed with each other that his soil was ideal for tobacco growing. Clearly, as early as the 1920s, competition between companies for the best growers in the most favourable areas had already begun.

Most farmers in the district were small holders making a bare living from a variety of crops including hops, pip-fruit, berries and melons. Some, like Frank Moss, also ran a few stock and grew a little wheat. The unpredictable markets for these products during years immediately following the First World War saw such farmers getting low returns from their land. By contrast, tobacco growers were said to be realising up to £60 per acre. This Depression of the early twenties had a particularly adverse effect on the outlying areas such as Stanley Brook where the land was marginal for horticultural crops. Farmers in Stanley Brook had relied mainly on sheep and cattle stock; now wool prices were below the cost of production and they had no hops or small fruit in their area. In Tapawera hops had been a major income-earner. By 1930 a local tobacco grower, F.W. Gibbs, was to remark that, where a few years before there had been fifty hop growers in his area, now there were three. The Australian hop market had been virtually closed to New Zealand hop growers and a small shipment sent to Dublin had received a very low price. In Motueka the picture was similar. Those who had previously been totally dependent on small fruits (such as red and black currants) and hops were just managing to exist. In the words of F.W. Gibbs, 'tobacco is the only thing that can save them'.[22]

So the attraction of a cash crop like tobacco was obvious in an area where holdings were so small 'you had to cultivate your back yard to make a living'.[23] The dramatic rise in the numbers of commercial growers of air-dried tobacco from approximately ten in 1922 to more than 100 in 1925-26,[24] with total growers numbering 160 by this time, illustrates the willingness of farmers to try out a crop which might improve their financial position. In valleys all over the region numbers of small farmers were adding tobacco to their range of crops. They found that, if satisfactorily grown under contract, the rewards were significantly above those for other farm produce. Cyprian Brereton described tobacco as 'the most reliable crop he ever grew' both in yield and in value.[25]

New growers moved into both air-dried tobacco growing and flue curing. In 1926, at the age of thirteen, Dudley Eggers began growing tobacco in Sunrise Valley, Upper Moutere. Grown under the direction of his father, Dud's first crop was about two acres of air-dried tobacco and was sold to the National Tobacco Company through Cecil Nash. The crop was packed in sacking bales resembling wool bales and was shipped to Napier through the Mapua wharf. One of Dud's early crops was spoilt in transit but the company paid out sixpence per pound to keep him interested in growing again the following season.[26] In Dovedale the contracts offered by Wills in 1927-28 were accepted by at least eight of the valley's small farmers. The largest acreage was planted by W.H. Burnett who grew four acres, with H.A. Thorn planting out three acres in his first year. By the following year, numbers in Dovedale had doubled and the sixteen growers planted over forty-six acres between them. Well-known local names involved included the Brislanes, Hawkes, Wins, Jordans and Davies, with C.A. Silcock added in 1929-30.[27]

In the Riwaka area, the more than forty air-dried tobacco growers were joined by at least ten growers of kiln-dried tobacco in 1927-28, and a further eleven the following season. Almost all of the now well-known Riwaka names appear in these early lists of growers – Rowling, Fry, Inglis, Jenkins, Hickmott, James and Askew were among the early Riwaka kiln growers. Similarly in Pangatotara, Woodstock and Upper and Lower Moutere farmers took the gamble on building kilns and associated buildings, and on committing a significant amount of farm time and labour to the cultivation of tobacco. At least three kilns were operating in the Moutere area in 1927-28 and a further five appeared in 1928-29. Here, noted family names in included Best, Hurley, Hodgkinson, Beuke and Bensemann.

A few farmers in the Upper Takaka and Sandy Bay-Marahau areas also ventured into the new crop, but with only moderate success. There was later to be much criticism of the way in which these growers were treated by Charles Lowe and the Department of Agriculture. Around Motueka itself the thirty-three growers of air-dried tobacco were joined by one kiln owner in 1927-28 and two more in 1928-29. As with other areas, growers now included all the familiar family names – Staples, Parkes, Atkins, Primmer, McGlashen, Eginton, Stephens and Inwood to name just a few.

By 1930 tobacco had become the most widely grown crop in the district. Despite some fluctuations in grower numbers, the basis of the industry had been laid down, methods were becoming refined and company affiliations established. Over the next few years the practical knowledge of the Motueka growers was to be complemented by a growing politicisation. The need to protect their interests saw the largely isolated, often fiercely independent rural small holders group together in a variety of attempts to secure the position of their industry.

NOTES

[1] T. Husheer, notes, p.7.
[2] National Archives file Tobacco Board LE 1/1 1932, list of manufacturers.
[3] Interview with Moss's daughter, Betty Hambleton, Motueka, 12.6.95.
[4] *Nelson Evening Mail*, 22.10.88.
[5] T. Husheer, notes, p.6. The figures of tobacco passing over the Nelson and Motueka wharves during the years 1921-24 cast an aura of doubt over the production of tobacco in the early years. Although it is consistently recorded that these growers were active during this time, port figures show no evidence of the tobacco being forwarded to the North Island manufacturer.
[6] Committee of Inquiry correspondence (1930) B7 (Nash).
[7] Brereton, C.P., *Vanguard of the South*, A.H. & A.W. Reed, Wellington, 1952, p.167.
[8] Minutes of Evidence, Committee of Inquiry into Tobacco Growing, 1930, p.283.
[9] Ibid, p.400.
[10] Ibid, p.3.
[11] This building is described as a small tin shed a little back off the road and was situated at 352 High Street next to the site presently occupied by Clark's Upholstery.
[12] Minutes of Evidence, Committee of Inquiry into Tobacco Growing, 1930, pp.645-46.
[13] Ibid, p.289.
[14] Committee of Inquiry correspondence, 1930, 27.8.27.
[15] Minutes of Evidence, Committee of Inquiry into Tobacco Growing, 1930, p.155.
[16] National Archives file LE 1/1 1930/16 History of the Industry B6
[17] Ibid, B7.
[18] Interview with Helm's son Herbert Helm, Pangatotara, 7.3.96.
[19] Waugh J.R., The Changing Distribution of Tobacco Growing in Waimea County, MA thesis in Geography, 1962, fig. 25.
[20] Minutes of Evidence, Committee of Inquiry into Tobacco Growing, 1930, p.482.
[21] Ibid, p.169.
[22] Ibid, pp.171, 217 and 218.
[23] Ibid, p.241.
[24] Waugh, op cit, figs 23 and 24.
[25] Minutes of Evidence, Committee of Inquiry into Tobacco Growing, 1930, p.167.
[26] Interview with D.A. (Dudley) Eggers, Richmond, 21.3.95.
[27] Minutes of Evidence, Committee of Inquiry into Tobacco Growing, 1930, lists of growers.

CHAPTER FOUR

How It Was Done

The process of cultivating and drying tobacco, both air-dried and flue-cured, in New Zealand was largely based on practices followed elsewhere in the world. The early advisers, except Cecil Nash, were experienced in tobacco culture overseas, especially in the US, China and Australia. They brought varying ideas to this country and local growers had to adapt their advice to suit the conditions in their particular area. The system of farm ownership in New Zealand differed from many tobacco-producing countries. In Australia, for example, much tobacco production was carried out on a share-farming basis. In African countries, large tobacco plantations were developed, with the labour being supplied by native Africans who lived on the plantation. In America, plantation land was often leased by poor whites and conditions were primitive by New Zealand standards.[1] The individual land ownership system in New Zealand meant that early tobacco growers in this country were more likely to carry out most of the work themselves, with the help of their families. For air-dried tobacco, the process of production was to change little, apart from advances in farming technology, over the succeeding years. Changes in flue curing were more dramatic, as kiln design, stoking procedures and leaf handling were hit by technological change.

Air-dried tobacco, commonly known as Burley, was the mainstay of the New Zealand tobacco industry until the late 1920s. Known as the non-fermentation method, air drying resulted in rich, golden-brown tobacco suitable for chewing and pipe smoking. Burley was relatively easy to grow and required little specialised equipment. At this time, the small areas planted required almost no labour apart from family members; in some cases several growers cooperated in the harvesting of their crops. Seed was provided by the company representatives and sown in prepared seed beds. Some of these beds were based on a square pit, six feet by six feet, edged with boards and filled with horse manure and soil. The rudimentary frame was covered with scrim to prevent too much sunlight as the seeds were germinating and beginning to grow. The scrim could be rolled back to allow some sun to harden the young plants ready for planting out into the fields. The plants were not thinned, as they were later, so one seed bed would provide a great number of plants.[2] Cecil Nash's directions for seed beds were that the bed should be ten feet by four feet, with an eight-inch-deep frame let into the ground about an inch. The frame should be filled with rich soil and levelled off to within four inches of the top. The seed was to be sown mixed with wood ash – half a teaspoon of seed being sufficient for each bed. After sowing, the bed was pressed down firmly with a board then the whole bed covered with scrim or buttercloth and kept covered until the plants were about two inches high.[3] Later, calico was used for seed-bed covers. The covers required hemming, and some women estimate they have hemmed up to 15,000 yards of calico, both making new covers and repairing old ones.

From the late 1920s plants were obtained in boxes from specialist nurseries, Broughs and Branfords of Nelson being the main suppliers. Boxes containing 1500 to 2000 seedlings were battened together and transported by truck to Motueka where they were portioned out to growers. The advantage to growers of this system was threefold: planting was brought forward as the plants were considerably ahead of farm-raised plants, the plants were hardier which meant less replanting, and the relative stability of plant numbers in the boxes gave growers a better idea of how many plants they actually had.[4] Planting or 'pricking out' the tiny seedlings into the prepared beds was a laborious task, tiny holes having to be made in the beds with a special implement and the delicate infant plants being dropped individually into them. With thousands of plants needed for each acre of tobacco, the individual planting of seedlings is recalled as one of the less popular tasks of the tobacco-growing routine. During the 1930s the prevalence of mosaic disease was linked to the extra handling of plants involved for nursery-raised plants and the practice of using farm-raised seedlings gradually prevailed.

After six to nine weeks, when about six or eight inches tall, the seedling plants were hardened off and were ready for planting out. The beds were well soaked with water for easier pulling of the plants, which were placed into boxes for carrying to the prepared fields for planting. For some years many growers planted out their fields by hand, a laborious process whereby each plant was dropped along the line of planting ahead of the planters, who formed a hole with a hand trowel, dropped the plant into it and smoothed the earth around it almost all in one motion. Over time many became highly proficient at hand-planting and could plant out a surprisingly large area in a day, often more than an acre.

Partly mechanised horse-drawn planters had been imported by Gerhard Husheer for his plantation in Hawke's Bay, and similar planters were in use by 1922 on the company plantations at Riverhead.[5] In appearance these planters were similar to a seed-drill. Behind the driver, two planters sat side by side, legs outstretched in front of them and a box of plants between their knees. The driver needed a good eye to keep a straight line and if the horse played up or the planters lost their rhythm the whole process collapsed in chaos. A rotating gadget on the machine drilled the holes, delivered a dollop of water from a tank beneath the driver and drew the soil around the plant. The process was tedious, but planters had to keep alert and maintain the rhythm of the rotating mechanism.[6]

The National Tobacco Company offered favourable financial arrangements for growers wishing to purchase these mechanised planters, the money being advanced interest free and repayable from the first payment for the crop. The young plants were planted at two-foot intervals in rows approximately three feet apart, leaving room for a sled to be drawn between rows during harvesting.[7] On some farms wires were strung across the paddocks to ensure the tobacco was planted in straight lines,[8] while others used a grid frame or chains attached to a yoke to mark out planting positions. Later Dovedale farmers used diagonal planting in the corners of their paddocks and horses were trained to head on the diagonal toward the end of the appropriate rows in order to create this mitred effect.[9]

Manure played a vital role in the cultivation of tobacco. While it was generally thought that tobacco could be produced on even the poorest soil, it was also acknowledged that the correct balance of manure dramatically affected the finished product. In early years a mixture of superphosphate, blood and bone, sulphate of ammonia and sulphate of potash was used.[10] Later, fertilisers specific to tobacco growing were produced. In the Motueka-Nelson area fertiliser mixtures were obtained from local freezing works for many years – more recently large fertiliser works such as Kempthorne Prosser and Ravensdown have produced these specific mixtures. Covering finance for the necessary manures was often provided by tobacco companies. Later, bulk buying through growers' organisations allowed savings for growers.

Cultivation of the crop during its growing season was essential to keep down weeds and to monitor the progress of the plants. Once nearing full height, the plants began to form flowerheads and in most cases these were removed, or topped. Topping was necessary to prevent flower petals from falling and 'burning' the leaves below. Suckers, or lateral growths from the leaf junctions, were also removed to promote growth in the main leaves. Hoeing, weeding, topping and suckering, or lateralling, were all tasks initially performed by family members, especially women and children, who were considered particularly suitable for such tasks. Women, according to their male partners and employers, liked the work as 'it was clean and healthy'[11] and they were reputedly faster than men at much of the work, especially suckering. Some maintained that children attained an expertise in these tasks which was rarely shown by adults.[12]

When they were mature, air-dried tobacco leaves showed signs of drooping and numerous yellowish spots appeared on the leaf surface. The leaves became brittle and a gummy substance was secreted which was sticky to the touch.[13] When this stage was reached, the whole tobacco plant was cut down and left on the ground to soften, thus preventing damage from handling while freshly cut and brittle. Later, leaves were harvested singly. The harvested tobacco was placed on stretchers, or into sacking bins, for cartage to the drying sheds. At least two workers were required to carry the stretcher or bin – later bins could be mounted on trestles for easier lifting. Horses-drawn carts, or sleds, were used to carry the crop from the field to the barn where the plants were hung over heavy sticks in pairs and then arranged in open-sided barns to dry. In some cases the barn was equipped with wires on which the plants were strung, having had their stems split for the purpose. Cecil Nash advised hanging the sticks twelve to fifteen inches apart for the first day, after which they could be moved to within eight inches of each other depending on the bulk of the plant.[14]

Curing air-dried leaf tobacco to the desired dark brown needed for satisfactory production of chewing and pipe tobacco required great care and fortitude. The leaf was vulnerable to damp conditions and some crops were moved into the open air during the day to aid in the process. Where the weather was stable, the sticks might be left outside for up to three weeks. Other growers lit fires under the drying leaves, using charcoal burners or sometimes using tobacco stalks for fuel. This was to keep the air dry and stave off mould, rather than to attempt to cure the tobacco. In later years air-dried tobacco was sometimes finished off in kilns, some as small as twelve feet square.

The crop could take up to six weeks to dry, which allowed drying sheds and barns to be used only once in a season.[15] Once dried, the leaves were stripped from the plants, roughly sorted or graded, usually by size and colour – early grades were 'lugs' (bottom leaves), 'firsts' and 'seconds' (lower and middle leaves). The leaf was then 'hanked' – bunches of fifteen to twenty leaves bound together with a leaf secured around the top to hold them. The leaf was then pressed, often, in the Motueka-Nelson district, using an existing hop press, and packed into large sacking bales similar to hop or wool bales. The bulk of the air-dried tobacco from the Motueka-Nelson district was shipped to the National Tobacco Company's bond store at Port Ahuriri.

Flue-cured, or kiln dried, tobacco, while trialled in 1919 by Charles Lowe, did not become widespread until after the 1926-27 season when W.D. & W.O. Wills began to establish contracts in the area, embarking on what was virtually an experimental programme during the years 1928-30. Flue-cured tobacco had the advantage for manufacturing companies that, besides producing the lemon-coloured leaf required for cigarettes, it had been taken one step further in the manufacturing process towards the finished product.

Cultivation and harvesting methods of flue-cured tobacco were similar to those for air-dried tobacco, although far greater areas could be planted and harvested with the quicker drying process. Areas for flue-cured tobacco ranged from five acres upwards, with some growers planting up to sixteen acres and building two or three kilns to cure their crop. Different varieties were used for flue curing and these varied through the decades. For flue curing the leaves of the plant were picked as they ripened, with pickers going over the crop several times to harvest the different types of leaf, each with its own peculiar description. Leaves at the bottom of the plant were known as 'lugs'. Above them were the best leaves – long, broad leaves called 'cutters'. The next leaves formed the bulk of the crop. Darker and narrower than cutters, these were known as 'first leaf' and 'second leaf'. At the top of the plant were the 'tips'.[16]

The harvested leaves were laid on the ground ready for teams of 'picker-uppers' to collect and load onto the wagons for transport to the sheds, where they were prepared for loading into the drying kilns. At the shed complexes, teams of workers were required to tie the leaves onto sticks and to load the sticks into the kiln. Thousands of tobacco sticks were needed for this purpose and were mainly of manuka, although some used white pine and others considered red birch made good sticks.[17] Procuring the sticks was a time-consuming job. Many had to go off-farm to secure sufficient sticks, and manuka sticks often needed to be stripped of bark and twigs. If carefully used, however, sticks could last years – some farmers boast of still having sticks on hand that could be up to fifty years old.

Tying tobacco onto the sticks became a specialised task, almost an art form. Performed almost entirely by women, tying involved 'passers' taking two or three leaves at a time from the bins of leaf brought in from the field and handing them to a 'tyer'. The tyer deftly wove each 'hand' of leaves onto the stick, first on one side then the other, securing the leaf with string which was looped expertly around each 'hand'. The speed reached by experienced tyers sometimes made it difficult for passers to keep up. Sharp words were often exchanged if the tyer's outstretched hand remained empty too long. Completed sticks consisted of between thirty-six and forty hands. When full, the sticks were lifted from the tying stand and passed into the kiln to be hung in tiers ready for curing. The usual aim was to fill a kiln (approximately 750 sticks) in one day to avoid deterioration in the green leaf prior to drying.

Early kilns were built to standard dimensions. Wills financed kilns which measured sixteen feet square at the base and stood twenty feet to the eaves. Use of materials varied slightly but most were constructed of corrugated iron with Polite interior lining. Some filled the cavity between with sawdust for insulation – this proved to be unsatisfactory as it got damp easily and was also very difficult to douse if it caught fire. Given the labour-intensive nature of work during the 1920s and 1930s, it was no joke for a grower who spent days hand-filling the walls of his kiln with sawdust only to find it an even longer job to remove it when such problems became clear.[18] In the early 1930s pumice was advertised as an alternative insulation material, although few now recall its use.

Early standard kilns were eight and a half racks high[19] and could cure approximately 1000 pounds of leaf at a time. A full kiln of tobacco required up to five days to cure. For many years kilns were fired by the up-draught method, with the heat entering the kiln below the tobacco and rising through natural convection to the kiln top or 'attic'. A furnace supplied heat through flue pipes on the floor of the kiln. The curing process consisted of four stages: 'colouring', which changed the green leaf to yellow; 'fixing the colour', which killed the leaf so that no further changes occurred; 'drying', which dried the bulk of the leaf; and 'drying the midrib', the drying out of the thick midrib of the leaf. Over the duration of curing, the temperature of the kiln was raised in stages to 170-180 degrees. Humidity was controlled through ventilators in the kiln tops.[20] Constant watchfulness was required, as was meticulous attention to the temperature. Charles Lowe detailed the shifts he worked with a grower when watching kilns. The two worked around the clock monitoring the curing process.[21] Coupled with the need for maintaining fuel supplies, this made kiln drying probably the most onerous and least enjoyable part of the pioneer tobacco farmer's ritual.

Kilns were initially wood-fired, with coal or coke being added to the range of fuels in the 1930s. Gathering wood for the furnace was another mammoth task. Many growers' sons recall hours spent gathering and cutting up the required fuel to keep a kiln fired for the necessary four or five days. Those with on-farm supplies were fortunate – willows along river banks were a favoured source. One second-generation grower in Dovedale recalls helping with wood-gathering at an early age. His job was to wait at the sheds for a horse to bring (unattended) a log or logs, which had been attached to its hauling chain by his father and older brother down by the river. He would unhook the timber and the horse would return, again unaccompanied, to the waiting wood-gatherers.[22] Splitting and stacking the wood consumed more hours, as did constantly monitoring the kiln fire, both to maintain the correct temperatures and to guard against the ever-present fire risk. Up-country growers knew well that a fire would destroy both their kiln and their crop, and possibly more besides, long before assistance could arrive. Photographs of an early Motueka kiln show a structure with a Polite base rising to about ten feet, with a wooden upper part of a further ten feet which was probably not insulated. Not surprisingly, only three kilns of tobacco were cured in this kiln before it burnt down.[23]

Concrete bases, approximately four feet high, were added to many kilns to reduce the risk of fire, and netting was often strung beneath the hanging leaves to prevent them from falling onto the hot flue pipes and catching alight. Monitoring kiln temperatures became probably the worst feature of life for those growers who moved into flue curing. For the five days of each kiln's curing time, little sleep was possible for the grower, his sons, or any trustworthy worker who could be tempted to keep this vigil. Most kilns had a bunkshed area where the furnace-watcher could rest, but alertness was always necessary and was maintained by various ingenious measures. One farmer slept with a spoon in his hand held over a bucket – when the spoon fell into the bucket it was time to check the kiln.[24] Gordon Rowling, of Riwaka, used an alarm clock system; the clock was set to ring every twenty minutes so that he could tend the fire regularly. He hated the clock so much that, when his system became conditioned to waking right on the twenty minutes, he set the clock for a few minutes beyond this. The clock seldom got the chance to annoy him.[25] Yet another grower kept watch sitting on a four-by-two beam. If he fell asleep, he fell off. One well-known Riwaka grower used physical discomfort to wake regularly – he would 'sleep' on the coal heap without his shirt on.[26] Many took weeks after the end of the harvesting season to get back to a normal sleeping pattern. Some recall falling asleep on the ground around sheds while being unable to sleep properly in bed at night. Lack of sleep made life stressful for many a tobacco grower's family. In the late 1920s some were already wondering if they had made the right choice in opting to build a tobacco kiln.[27]

Once satisfactorily dried, the sticks of leaf were unloaded from the kiln, often during the night in order that the kiln would be emptied ready for the next round of harvesting. The dried leaf was then bulked down. This process, which assisted in the maturation of the leaf, involved stacking the sticks of tobacco in an orderly fashion. Some were more orderly than others – occasionally stacks could be seen tilting ominously. Bulksheds could be filled to the ceiling for up to three months. When harvesting was completed the dried leaf was taken from the bulkshed, steamed lightly for easier handling, then graded leaf by leaf according to size, colour and company requirements. Grading became one of the major winter occupations for the tobacco-growing family, and later for local women and summer workers staying on in the district. Depending on the size of the crop, weeks at a time were spent in the tobacco-laden atmosphere of the grading shed. Before buying sheds were established in Motueka, the roughly graded leaf was packed in bales weighing 300-400 pounds, trucked to the wharves at Motueka and Nelson and shipped to the processing factory, which was then likely to be Wills' factory at Petone. In later years all manufacturing companies in New Zealand purchased kiln-dried tobacco, although small quantities of air-dried tobacco continued to by grown into the 1980s.

Early freight charges were considered exorbitant at £1 per bale. Shippers were apparently convinced that tobacco was a very dangerous cargo,[28] possibly by those who considered dried tobacco to be highly combustible. Later, when Cecil Nash persuaded the shippers that the industry would grow to become a most important feature of the local economy, the freight was reduced to a halfpenny per pound. With bales weighing around 300 pounds, this equated to approximately 12s 6d per bale. During the 1920s the ports of Motueka, Mapua and Nelson were used as points of exit for this tobacco. Growers were at the mercy of the shippers for the correct treatment and stowage of their produce; some consignments were spoilt through being stowed under other freight such as boxes of apples.

Gradually companies set up buying floors in Motueka so that the graded leaf could be assessed and purchased locally. The buying season generally commenced immediately after Easter. A field officer (Ian Hamilton for Wills in the 1930s) went around the farms to set the grades suitable to each crop. As sufficient leaf was graded, usually by mid-May, buying days were arranged so that each farmer would bring a load of about 1000 pounds of leaf to the buying shed. Company officers were on hand to inspect the leaf as it arrived. Early inspections at Wills' receiving shed were carried out in the cart dock entrance where the natural light was particularly good. The buying officer assigned a price to each bale, although the farmer could bargain if he wished. Once the price was settled, the leaf was weighed and removed into the shed. After baling, the leaf was shipped to the manufacturing factory where it was opened, and regraded if necessary.

Companies advised purchasing leaf as soon as possible. Unless carefully looked after the leaf would deteriorate, especially if moisture remained within the leaves or bundles. Green tobacco could improve with keeping. Manufacturers often purchased green leaf at reduced prices but were not in favour of increased production of such leaf. Generally, both companies and the government discouraged the practice of having quantities of tobacco 'lying about in growers' sheds'.[29] Such leaf would only encourage cut-price selling and smuggling, both of which were relatively commonplace up to the mid-1930s.

Once the crop was completely harvested, it was necessary to plough under the plant stalks. If left in the field these could harbour diseases and pests which could severely affect the next crop and the crops of other growers. The whole process, including grading and selling, was usually completed by July-August each year. Some were able to take a little time off before beginning all over again.

Notes
[1] From autobiographical notes by Charles Lowe, p.2, supplied by R. Lowe, Wellington.
[2] Interview with Pat Beatson, Orinoco, 3.8.95.
[3] Minutes of Evidence, Committee of Inquiry into Tobacco Growing, 1930, p.134.
[4] Biographical papers of Gordon Rowling, supplied by Sally Goodall, Riwaka.

[5] *New Zealand Tobacco Farm*, p.2, publisher and date unknown. Copies supplied by J. Husheer.
[6] C.B. (Pat) Beatson, *The River, Stump and Raspberry Garden*, Nikau Press, Nelson, 1992, p.82.
[7] Op cit.
[8] Interview with former grower R.J. (Bob) Williams, West Bank, Motueka, 15.6.95.
[9] Interview with grower George Douglas, Dovedale, 18.7.95.
[10] *New Zealand Tobacco Farm*, p.2.
[11] Minutes of Evidence, Committee of Inquiry into Tobacco Growing, 1930, p.295.
[12] *New Zealand Tobacco Farm*, p.2
[13] Op cit.
[14] Minutes of Evidence, Committee of Inquiry into Tobacco Growing, 1930, p.136.
[15] Waugh, J.R., The Changing Distribution of Tobacco Growing in Waimea County, MA thesis in geography, 1962, p.36
[16] R. Thomson, *Flue-cured Tobacco Growing in New Zealand*, Department of Scientific and Industrial Research, 1948, p.40.
[17] Interview with growers George Douglas and Hugh Burnett, Dovedale, 18.7.95.
[18] Ibid.
[19] There were variations in kiln sizes, with some kilns being seven and a half racks high. Converted hop kilns also varied in dimensions and the racks were usually further apart, making loading difficult for workers with short legs. Interview with Hugh Burnett.
[20] Thomson, op cit, pp.34-36.
[21] Personal papers of Charles Lowe.
[22] Interview with Hugh Burnett.
[23] Personal papers of Charles Lowe.
[24] Interview with former grower N.J. (Pat) Martin, Motueka, 12.6.95.
[25] Notes of Gordon Rowling.
[26] Interview with former grower John Stevens, Ruby Bay, 6.12.95.
[27] Minutes of Evidence, Committee of Inquiry into Tobacco Growing, 1930, p.247.
[28] Brereton, C.P., Vanguard of the South, A.H. & A.W. Reed, Wellington, 1952, p.168.
[29] Notes by R.B. Smith, W.D. & W.O. Wills, 9.9.36, National Archives File Tobacco Board 1/1.

CHAPTER FIVE

Companies and Controversy

The entry of Wills into the contract growing of tobacco in New Zealand was given wide publicity in the media. Reports suggested that prices offered by Wills for the Motueka Growers' Association crop in 1927 were as high as 2s 6d per pound. This sparked interest in tobacco growing in other parts of New Zealand and led to a rash of companies seeking to capitalise on the predicted profitability. Around the country ventures began in the hope of competing with the Nelson growers. At least eleven companies were set up in the Auckland region between 1928 and 1930, almost all with the idea of plantation growing of tobacco. By the 1929-30 season there were three growing companies operating between the Bay of Plenty and Northland with a combined area of 370 acres. The following year, eight companies were predicting a combined planted area of more than 6000 acres.[1]

The formation of these companies was surrounded with controversy. Some suspicion lingered that the Department of Agriculture was favouring certain commercial interests, especially those proposing tobacco growing for export. Some company claims of the profits to be made from tobacco were based on the returns said to be obtained by Nelson growers – some included reports written by Lowe.[2] The transfer of Charles Lowe to Auckland during the 1927-28 season lent fuel to the claim that he was working to promote the interests of tobacco-growing companies. The department said Lowe's transfer was to meet requests for advice from Northland growers, and that Wills had their own instructors in the Motueka district.

There had been some friction between Lowe and these advisers. Lowe had received a letter from Whittaker, Wills' instructor, to the effect that he was not to comment when visiting their kilns and was to refer always to the then company instructor, Sedrick Brame, as being in charge of curing.[3] Lowe was apt to speak his mind when he saw fit and was even said to intervene if a kiln was not being built in what he considered to be a proper fashion.[4] On his transfer, Charles Lowe put his efforts into advising groups of growers on the suitability of their land for tobacco growing and, if they decided to go ahead, on methods of cultivation and curing. Much controversy surrounded these activities and Lowe was to suffer from this for many years.

Following suggestions from two prominent Northlanders, Captain Allen Bell, MP, and Dr Smythe, that tobacco growing might be advantageous to Maori farmers in the Hokianga area, Lowe visited the district in 1927. Subsequently numbers of farmers in the area, including both Maori farmers and European settlers, set about growing tobacco. Similarly, growers in Te Atatu (an area where Gerhard Husheer had previously purchased tobacco) requested Lowe to address them and formed a group to test out the crop. On leaving the area, Gerhard Husheer had left behind many of the Auckland syndicate's assets. These included planting machines and buildings at the Tobacco Growing Centre at Te Atatu, which were given for communal use.[5] In the Bay of Plenty Lowe advised a group of Maori growers at Rotorua through the Arawa Trust, chaired by Major Rodger Dansey. In Marlborough a group of growers set up trial crops in 1928, later claiming that Charles Lowe had made extensive promises on the profitability of export tobacco.

In 1928 there were fifty-three growers in the Auckland area, including the Hokianga and Rotorua growers as well as plantations in Auckland and Tauranga. Besides the 559 Motueka-Nelson growers, in the south there were twenty Marlborough growers. Almost all of the ventures outside the Motueka-Nelson district ended dismally, often bringing financial distress. On a visit to the Hokianga in 1929 Lowe found that the mainly Maori growers, while keen and possessing ability, lacked the correct information on methods of drying and storing their tobacco. Lowe described the Hokianga drying barns as being thatched with nikau and with nikau sides which encouraged mould, especially in the higher humidity of the north. The resulting tobacco was of patchy quality. In the Kohu Kohu, Waima and Omanaia areas he found quantities of good leaf but in other areas, such as Whirinaki, the farmers were floundering.[6] Lowe observed model conditions, especially for the Maori women workers, in the Hokianga, but the area needed proper curing barns in order to produce consistently good tobacco. Between forty growers in this area there was only one kiln.[7] Despite this, Lowe estimated that there were approximately ten tons of tobacco leaf available for sale. The Hokianga crop, however, met with failure as there proved to be no market for this leaf. Finally the government offered a payment of eightpence per pound from which expenses of transport and storage were to be deducted.[8]

The Hokianga crop was shipped to Wills' factory in Petone where Charles Lowe supervised the redrying of the leaf. Subsequently he attempted to sell it to both Wills and National. Neither company was interested, Wills because it purchased very little air-dried leaf, and National because it felt the quality to be well below any tobacco it could use.[9] In 1930 the Hokianga crop remained unsold, although a large proportion of this leaf (14,000 pounds) had been sent to London in an attempt to dispose of it on the international market. This venture was unsuccessful, with prices as low as one penny per pound being offered.[10] The growers of the Hokianga were sufficiently discouraged to make no further plantings of tobacco.

The Arawa Trust growers of Rotorua fared a little better, although the results were not nearly as good as they had hoped. Arawa growers, mainly returned soldiers and their wives,[11] planted fifty acres in plantations at Kouto, Awhatu, Tengi and Morea. Contemporary photographs show very orderly plantations divided into sections by maize windbreaks. From these images the crop appears good, with Charles Lowe's advice being followed efficiently. However, having just two kilns between twenty-five growers, they produced a mahogany crop of tobacco which was suitable for pipe tobacco only. For a time no market could be found for this tobacco. Eventually the crop was bought by the National Tobacco Company at prices ranging between sixpence and 1s 9d per pound.¹² Cecil Nash claimed that Dansey and the Rotorua Maori had been misled by Lowe. Although Gerhard Husheer purchased the crop, he said he did so more as a favour to the Maori and hoped the experience would teach the growers a lesson. Dansey was more direct in his criticism, referring pointedly to those 'irresponsible Pakeha' who had fostered fictitious values for tobacco and unrealistic expectations amongst the Maori growers.¹³

The eight Te Atatu growers felt particularly aggrieved. There proved to be no market for their tobacco at all, despite Lowe having visited their properties, inspected their crop and advised them that they would receive a shilling more per pound if they were to flue-cure the leaf. Four growers borrowed £100 to build a kiln and proceeded to cure the tobacco without instruction. The results were devastating. By 1930 the growers had received nothing for their crop. Subsequently some of their leaf was disposed of, at great loss, with £46 15s 4d being received for flue-cured tobacco that had cost £269 5s 8d to produce. Those Te Atatu growers who stayed with air drying were little better off. One received £6 8s 6d for leaf costing £74 18s 2d to produce, while another received 7s 3d against production costs of £36 0s 5d. Criticism was again levelled at Charles Lowe, who was said to have encouraged this group but then left them with little instruction and infrequent visits.¹⁴ Their tobacco was ultimately forwarded to Wills' factory in Petone where it was to be prepared for export, but little further was heard by the growers.

The other North Island growers during this period were companies formed to grow tobacco on a large scale. Much of this activity was purely speculative as none of the companies appear to have secured contracts for the sale of their tobacco to manufacturers. Most, in fact, relied on the government drive to establish an export trade in New Zealand tobacco. Often the use of land largely unsuitable for tobacco growing saw thousands of pounds lost in risky ventures. Some companies were formed through the issue of bonds, such as the Empire Tobacco Corporation, which issued bonds at £30 each.¹⁵ This form of financing new companies was not widespread in New Zealand at this time and was considered to be more appropriate for long-term ventures such as forestry. Although the issue of tobacco bonds had been criticised in Australia through the influential *Smith's Weekly*, it had been defended by the statement that investors' interests were 'protected by solicitors'.¹⁶ The proposed Empire Corporation issued both a comprehensive written prospectus and a glossy pictorial brochure featuring tobacco growing in many parts of New Zealand, including the Arawa plantations at Rotorua and private farms in Riwaka.

By 1930 the subscribed capital of companies in the Auckland area was almost £350,000. Production, however, did not keep pace with company claims. The Tauranga Citrus and Tobacco Company produced 6000 pounds from seven and a half acres at Papamoa in 1929-30 and planned to increase this to fifty acres. This company later formed the United Tobacco Corporation (Tauranga) Ltd which then proposed to plant 2000 acres. The New Zealand Tobacco Company produced 120,000 pounds of tobacco from 195 acres at Riverhead in 1929-30 and planned to plant 300 acres in the next season. In Kumeu, the Standard Tobacco Company planned a planting of 500 acres. At Shelly Beach, Helensville, the Tobacco Producers' Trust initially planted fifty acres and employed sixteen people; later this was planned to rise to 1000 acres. In the Bay of Islands, the Pacific Tobacco Company, formed by Australian interests, was planning to plant 1000 acres. In addition, Tobacco Growers (NZ) Ltd was growing 100 acres at Te Hana, Wellsford, employing twenty men and planning to increase its planting to 500 acres. This company intended to sell its production to an English manufacturing company.[17] None of these ambitious planting targets was achieved.

Company speculation was rife and newspaper articles condemned the scramble for quick profits. Gordon Rowling, of Riwaka, who had been involved in growing tobacco since 1925, had a position at Riverhead at the time (probably with the New Zealand Tobacco Company). He described how some of the speculative tobacco companies were operating:

> … six men formed an investment company, and by paying a small deposit they took an option on a tract of land. On the first farm started they held an option to buy at £25 per acre. When they floated the Company to grow tobacco, they, being the directors of the same Company, purchased this land at £120 per acre from their investment Company… The most flagrant deal I heard of was at Riverhead where they held an option on some undeveloped land at £10 per acre and then used all or nearly all the floated Company's monies to buy it (from themselves) at £180 per acre. Eventually they were discovered and one at least committed suicide.[18]

In 1930 the New Zealand Tobacco Growers' Federation (not related to the later federation formed by Motueka growers in 1937) was established. It was an association of tobacco-growing companies intended to operate in the manner of dairy cooperatives. The Department of Agriculture, through the Director of Horticulture, Campbell, and Charles Lowe, assisted with the formation of this organisation, which was designed for large companies. Even so, Campbell believed small growers could form an association which could then join the federation. As representation was to be based on acreage, associations of small growers could ultimately dominate the federation.[19] Campbell claimed he and Lowe had encouraged Motueka-Nelson growers to form an association, but this did not eventuate – an early opportunity to create strength and unity amongst tobacco growers was lost.

The federation of growing companies aimed to further the interests of growers, to repel whatever was prejudicial to tobacco growing, to undertake overseas marketing and to secure an adequate domestic market for New Zealand-grown tobacco leaf. Its members saw three main problems to be addressed: the difference in leaf quality and type between districts, persistent overproduction and the ability to carry over production to cover shortfalls in poor seasons.[20] Few of these objectives were met and some remained problematic to the industry right up to its deregulation in the 1980s.

While many of the new growing companies hoped to use existing manufacturers for the processing of their leaf, this hope was rarely fulfilled. In 1930 the production of these companies remained unsold; indeed this was still largely the case in 1934. Of fourteen companies directly involved with tobacco only four remained in 1934, the rest having been struck off the company register. Quantities of surplus tobacco were later sold for knock-down prices, bringing little, if any, return to investors. But speculation was not limited to companies. Most individual growers in the Motueka-Nelson area had contracts with one or other of the main companies or their agents, but some grew un-contracted crops in the hope of selling once the harvest was over. Charles Lowe estimated that there were at least sixty acres of speculative growing in the Motueka-Nelson area in 1930.[21] The resulting oversupply caused problems for both growers and companies and the industry was often described as chaotic.

For Marlborough growers the story was as depressing as it had been for their northern counterparts. Growers requested Department of Agriculture advice and permission to trial tobacco growing. A.E. Bartlett, the secretary of the group, claimed that Lowe had advised them to grow large quantities, telling them that Wills had erected a facility covering eight acres, that growers could not go wrong and that soon they would be exporting tobacco by the million pounds' worth. The twenty growers grew for one season but there was no market for their crop and the trial ceased. Bartlett admitted the department had issued warnings on the dangers of growing tobacco without firm contracts but the growers had 'unwisely followed Lowe's advice'. Cecil Nash was also critical of Lowe's dealings with growers in Marlborough, as well as those in the North Island, Sandy Bay and Takaka, claiming that Lowe had got growers started then left them to flounder just at the time they needed specialist assistance.

In fairness to Lowe, however, the Marlborough growers had asked to undertake the trials, but had then gone ahead without the support of the government guarantee scheme, and Bartlett himself appears to have been something of an opportunist. He had written a letter for publication regarding tobacco growing in New Zealand, saying he had had numerous enquiries from prospective growers, brokers and investors, but he later admitted knowing very little about the process of producing tobacco, its cultivation or handling. Bartlett's own interest in tobacco was apparently in his capacity as a nurseryman and he was primarily hopeful of providing seedling tobacco plants to growers. Cecil Nash advised him that growers could, in fact, easily grow their own.

All the same, these Marlborough growers felt let down. In the event, it was the National Tobacco Company which bought the bulk of this crop a year later at one shilling per pound higher than the government-guaranteed price. Wills purchased the balance at sixpence per pound over the government price.[22] Charles Lowe was convinced this was because neither company wanted to encourage the growing of tobacco for export. Although this is unsubstantiated, it is probable that Nash's feeling on this, that companies would then be offered only second-grade local tobacco, was shared by company management. It was also true that prices being obtained on the overseas market would have been uneconomic for New Zealand growers.

This whole chapter of events from Hokianga to Blenheim placed Charles Lowe in an increasingly difficult position. As a government-appointed instructor in tobacco culture he was obliged to carry out departmental policy, which at this time was to give advice on land suitability for a range of agricultural purposes including grain growing, sheep or dairy farming and fruit orchards. Tobacco was seen as just another horticultural option and the department advised accordingly.[23] While it was acknowledged that Lowe's 'tongue may have run away with him'[24] regarding the financial possibilities of tobacco, his enthusiasm was nevertheless defended by his superiors.

The Department of Agriculture was committed to the concept of exporting tobacco, and much of Lowe's time was spent in fostering this concept. Many of the companies set up to grow tobacco relied on technical advice from Lowe and this rebounded on him later, setting him at odds with other leading figures in the industry, such as Husheer and Nash. Lowe himself denied any links with particular commercial interests, but suspicion lingered that he used his position to the advantage of some. Here again the department's position proved awkward for Charles Lowe; while advising caution, the Director of Horticulture, Campbell, felt that companies could provide the necessary organised control for the industry.[25] In addition, the department continued to provide tobacco seed to almost anyone who requested it.[26]

Charles Lowe resisted all allegations and criticisms. He maintained that his sole focus was the promotion of successful tobacco growing wherever conditions were suitable. Part of his reasoning in establishing growers in many different areas was the desire to gather knowledge on the suitability of each location. Aware that this knowledge, along with much other research, was lacking, Lowe expressed a desire to see widespread tobacco culture throughout New Zealand.[27] It is easy to see how this desire led him to foster hopes in areas where success was virtually impossible. Lowe claimed (in 1930) that he had tried to discourage many of those keen to enter the industry as the markets were so uncertain. Certainly, despite the series of unfortunate results, his technical advice was always accepted and valued by would-be growers and he formed lasting friendships with many in the industry. A keen hunter, he was a welcome visitor on the Gibbons farm in the Pongakawa Valley where he enjoyed the opportunity of a rabbit shoot.[28] But Lowe's history, with its element of entrepreneurialism and self-teaching, stood against him in the minds of some in the industry who were unconvinced that his technical expertise and knowledge were sound. Among his severest critics were company representatives from both Wills and National.

For their part, growers were also often suspicious of manufacturers, believing that misleading, or less than complete, reports were sometimes given on the condition of their tobacco on its arrival at company facilities. The lack of local buying sheds was especially significant in fostering this belief. Many also felt that the price remained the same regardless of the state of tobacco on its arrival. Many stories are told of letters received by growers saying 'your tobacco arrived in a very poor condition and the price offered will be 1s 6d per pound' or 'your tobacco arrived in very good condition and the price offered will be 1s 6d per pound'. One disaffected grower, John Martin, took ship to Napier to check on his tobacco: having been offered sixpence per pound 'for mercy's sake',[29] he arrived to find the leaf being processed. Others claimed that the companies were manipulating contracts to suit themselves. It was said that the National Tobacco Company had a waiting list of 200 farmers wishing to grow tobacco and that, in the years when this company was the only buyer of local leaf, contracts were highly variable and unreliable, often changing from year to year. Billy Owen stated in 1930 that a grower might be given a contract for two or three acres one year, then one or none the following year.[30]

Despite these negative aspects, the reports of returns of £100 per acre sounded good to farmers struggling with the effects of falling or unpredictable prices for other rural produce. Grower numbers rose steadily. Cecil Nash stated in 1930 that at least twenty-nine farmers had got out of financial difficulties through growing tobacco.[31] But even at this early stage many were aware that the success of the tobacco farmers of the Nelson area stood on the use of unpaid family labour on the many small farms in the district. Billy Owen, by then managing a plantation in the Bay of Plenty, commented in 1930 that 'everybody knows that mother and children get very little out of it'.[32] C.P. Lock of the Tauranga Citrus and Tobacco Company agreed with this view of the basis of the success of tobacco growing in the Nelson area.[33] Not having to pay his family meant the male farmer appeared to make a living from his crops; consequently the returns appeared far more attractive than they really were.

Many farmers in the area *were* doing better than they had in the past. While the three-year Wills contracts were in force, the position remained relatively stable. However with changes about to be made in tariff provision and subsequent cuts in contract area, tobacco growers were hit by the first of the many shock-waves which were to become familiar to the industry.

NOTES

[1] Minutes of Evidence, Committee of Inquiry into Tobacco Growing, 1930, pp.10-11.
[2] Ibid, p.304.
[3] Ibid, p.293,
[4] Correspondence, Committee of Inquiry into Tobacco Growing, 1930, 31.1.28.
[5] T. Husheer, notes, p.6.
[6] Personal papers of Charles Lowe.
[7] Minutes of Evidence, Committee of Inquiry into Tobacco Growing, 1930, p.61.
[8] Report of the Committee of Inquiry. AJHR 1930 I-17 p.12.
[9] Minutes of Evidence, Committee of Inquiry into Tobacco Growing, 1930, p.146.
[10] Parliamentary Debates 1930, p.1001. George Black, MP for Motueka and chairman of the Committee of Inquiry.
[11] Brochure published by the Empire Tobacco Company, c1930, p.9.
[12] Minutes of Evidence, Committee of Inquiry into Tobacco Growing, 1930, pp.145-46.
[13] Correspondence, Committee of Inquiry into Tobacco Growing, 1930, 8.10.29 and 16.10.30.
[14] Ibid, 7.10.29
[15] Minutes of Evidence, Committee of Inquiry into Tobacco Growing, 1930, p.11.
[16] Ibid, p.19.
[17] All information in this section from the Minutes of Evidence, Committee of Inquiry into Tobacco Growing, 1930, pp.10-11.
[18] Rowling, notes.
[19] Minutes of Evidence, Committee of Inquiry into Tobacco Growing, 1930, pp.22, 357.
[20] Correspondence, Committee of Inquiry into Tobacco Growing, 1930
[21] Minutes of Evidence, Committee of Inquiry into Tobacco Growing, 1930, p.74.
[22] All material on Marlborough episode from Minutes of Evidence, Committee of Inquiry into Tobacco Growing, 1930, pp.73, 146, 200, 209, 404.
[23] Ibid, p.4.

[24] Ibid, p.618.
[25] Ibid, p.296.
[26] Correspondence, Committee of Inquiry into Tobacco Growing, 1930, B7, Nash.
[27] Lowe, quoted in Empire Tobacco Company brochure, c. 1930.
[28] Interview with Lindsay Gibbons, Pongakawa, 30.9.95.
[29] Interview with Pat Martin, 16.1.96.
[30] Minutes of Evidence, Committee of Inquiry into Tobacco Growing, 1930, p.36.
[31] Ibid, pp.125-48.
[32] Ibid, p.36.
[33] Ibid, p.42.

CHAPTER SIX

Trying Times and Trips to Wellington

Following the 1929-30 season, Wills reassessed its operations and requirements. Acreages had been reduced in that season as contracts for air-dried tobacco were not renewed; now flue-curing contracts were under review. Having taken almost all of the flue-cured tobacco produced by their growers from the first two years of their contracts, Wills found it necessary to reject considerable amounts of the 1929-30 crop. Although the amounts varied considerably, only a handful of kiln growers throughout the district had their entire crop accepted. Motueka grower C.L. Harvey was hardest hit, with 54 per cent of his tobacco rejected. C.E. Jordan of Dovedale had 39 per cent of his leaf rejected and W.M. Hawkes, also of Dovedale, returned home with 37 per cent of his crop. Pokororo was particularly badly hit as, besides J.T. Heath being left with 45 per cent unsold, Cyril Heath received nothing for 40 per cent of his crop and W.S. Win and Son had 37 per cent of their leaf rejected.

In all, more than 50,000 pounds of flue-cured leaf was unsold from the year's harvest. This was slightly over 17 per cent of the total kiln-dried crop of 293,812 pounds.[1] In a move to cull out growers who appeared to be producing an unsatisfactory product, Wills delayed offering contracts for the 1930-31 season. Then, on 22 July 1930, the company advised its Nelson agent, Buxtons, of a reduced list of 'approved growers' who would be offered contracts for the coming year.[2] The lists, or rather the omissions, produced a storm of protest from growers who maintained the company had given them nothing but praise for the quality of their tobacco and, above all, had encouraged them to build expensive kilns and sheds. With no tobacco contracts, these would now be completely useless. B.W. Jenkins, of Riwaka, had built a double kiln on company advice and was said to occupy some of the best land in the area, yet he was offered no contract for the 1930-31 season. A similar experience was related by John Duncan, of Riwaka. Sedrick Brame commented that if Duncan was to be cut out then 'all Riwaka growers should be cut out as he has some of the best land in Riwaka'. In Upper Moutere R. Hodgkinson felt he had been 'left high and dry', and in Dovedale James Balck was first advised in writing by Buxtons that he had a contract for the coming season, but then omitted from the Wills list. Balck wrote directly to Wills and was granted a contract for 3000 pounds.[3] In effect the Wills lists reduced the numbers of kiln growers from seventy-six in 1929-30 to thirty-six in 1930-31, although later amendments brought numbers up to over forty. The many Wills-financed kilns, now standing unused around the district, became known locally as 'Wills monuments'.

The worst-affected areas were Waimea and Upper and Lower Moutere, and only

Bernard Lusty of Richmond retained a contract for the coming season. These areas had produced over 50,000 pounds of flue-cured leaf in the previous year – the one remaining contract was for 3000 pounds. Other major producers in this area to miss out on a contract besides Hodgkinson were Daniel Hurley, Alex Best, George Perry, S.E. Burnett, L.K. Drummond and C.M. Bensemann. In Riwaka ten growers were left without contracts, while in Dovedale only nine of the original eighteen kiln growers were offered contracts. Wills planned to restrict their requirement for the 1930-31 season to 150,000 pounds. The lists of approved growers allowed for contracts of 144,200 pounds, with 800 pounds 'to be allotted preferably to Up the Valley or Dovedale growers'.[4] The remaining 5000 pounds would presumably be made up from any excess production from these list growers. In addition, the new contracts were to be let on a poundage basis, replacing the previous contracts based on acreage. With the variation in company requirements, many growers felt it would be necessary to plant up to double the normal acreage in order to produce the contracted leaf weight. As Wills had paid less than the contract price for two out of the previous three years, growers were faced with increased production costs and diminishing returns.[5] National Tobacco Company growers remained relatively unscathed by these cuts in contract areas, but the company believed that the New Zealand growing industry was near its limit. Growers rejected by Wills were therefore unlikely to be offered contracts by National.

These issues, compounded by proposed new tariff structures, generated sufficient heat for a government Select Committee of Inquiry to be set up in 1930. George Black, the then member of Parliament for Reefton, chaired the committee of twelve, which received evidence and submissions from a wide range of those involved in the industry. The terms of reference for this inquiry were essentially to investigate the effects of the introduction of new tariff structures on the industry as a whole. The committee held twenty-four meetings and members travelled to Motueka, Riverhead, Napier and Petone for a first-hand look at both growing and manufacturing. Petitions were received from growers from all parts of the Nelson province as well as submissions from the Te Atatu and Marlborough growers. At least seventeen growers attended the hearings, as did field experts Charles Lowe, Cecil Nash and Sedrick Brame, as well as company management and agents.

The committee looked at all aspects of the industry from its inception in an attempt to identify the causes of the existing problems. In support of their petitions those growers who travelled to Wellington gave graphic descriptions of the reality on their farms. In addition, enormous effort was expended on producing information on the previous three seasons: acreages, investment in kilns, ancillary buildings and equipment, production figures (including tobacco rejected in the 1929-30 season), numbers of employees, cost of production and projected available acreage were all recorded for each grower in each area. This was a major project given the resources of the growers and the communication difficulties at the time. Companies were less organised in their approach to the inquiry, possibly as they had far less to lose under the prevailing conditions. During the course of the inquiry the committee requested various company representatives to appear more than once to clarify specific points. The National

Tobacco Company, largely represented by Cecil Nash owing to the illness of Gerhard Husheer, relied on its history of support for its growers and what it saw as its battle with 'the Trust', as Nash consistently called W.D. & W.O. Wills. Convinced that Wills had vested interests in tobacco growing in other countries, he claimed that if 'they could smash the National Tobacco Company tomorrow they would do so'.[6]

Through its manager, R.B. Smith, Wills maintained that companies were learning about New Zealand tobacco just as growers were.[7] Having offered contracts for one to three years, something unusual in other horticultural sectors, Wills had reviewed the position in 1930 and now needed a greater supply of bright leaf. To achieve this it now proposed to allocate contracts on the basis of previous quality and ability to produce the desired leaf type. Holding fast to the opinion that New Zealand tobacco was inferior in quality to its overseas counterparts, despite the favourable report from their its London agent, Wills sought to restrict the overproduction of leaf unsuitable for cigarette manufacture. The new contract, therefore, included a clause requiring growers to deliver to the company receiving shed one pound of light-bodied leaf for every two pounds of heavy-bodied leaf. The company reserved the right to reject all heavy-bodied leaf in excess of this ratio even if the total weight fell within the amount contracted for.[8] Growers in Dovedale were confident they could easily meet this ratio[9] but those in less favoured areas found their contracts reduced or non-existent. Having found a crop which produced predictable cash returns, farmers in marginal areas were reluctant to give it up without resistance.

And so they came. Some, like Stanley Fry of Motueka, represented as few as three growers, while J.A. McGlashen of Riwaka spoke on behalf of 105 growers with 357 acres of tobacco between them. In all, more than 400 Motueka-Nelson growers were represented, all protesting at the lack of stability in the industry and fearing that tougher company buying policies and the proposed tariffs would lead to the complete demolition of their new livelihood. Companies, on the other hand, had diverse attitudes and these were clearly evident in the submissions they made. National Tobacco retained its commitment to domestic growers since these provided the basis of its manufacturing process and the appeal of its products to the smoking public. Wills, having offered the three-year contracts on a partly experimental basis, was far less committed to the domestic production of tobacco. Although it had financed kilns and other equipment, the loans were largely repaid and the question of surplus buildings was hardly the company's problem.

The select committee had to pick its way between these diverse viewpoints and did so with great depth of understanding. Members identified the difficulties of New Zealand growers competing with tobacco landed under favourable tariff arrangements. Without the supporting research and technical assistance which would enable New Zealand growers to understand the exact requirements of the companies, local growers were further disadvantaged. The way in which manufacturing companies appeared to be using the tariff structure and leaf definition for their own advantage came under close scrutiny and detailed questioning. The new tariffs provided for higher duty on imported tobacco if a company used no New Zealand leaf, but there was no incentive for manufacturers to use local leaf unless their products contained more than a third of the domestic leaf.[10]

A great debate began over the definition of 'manufactured tobacco'. There seemed to be no clear-cut opinion on just when the manufacturing process began. Some, like Fred O. Hamilton of Buxtons, believed that the drying process carried out by growers in their kilns was the first step in the manufacture of tobacco. Others maintained that even though the tobacco imported from the US had been cured, reconditioned and often stripped from the stem, it was nevertheless still unmanufactured tobacco. Since this attracted less duty there was an obvious advantage for companies in holding to this opinion. They benefited both from the lower duty, and from partly processed imported leaf being several steps ahead of locally produced tobacco in the processing procedure.

Smith, of Wills, agreed that 'strips' were indeed manufactured tobacco but that Wills was importing these as unmanufactured tobacco.[11] The Comptroller of Customs, E.W. Good, considered that stemmed tobacco (stripped from the stems) was manufactured tobacco and that it was a big concession to Wills to treat it as unmanufactured leaf and reduce duty accordingly.[12] In addition to this difficulty of definition, there were other means by which companies could avoid or manipulate the payment of tariffs. One of these was revealed by C. Lock, of the Tauranga Citrus and Tobacco Company. He maintained that tobacco with thirty cuts to the inch, suitable for cigarette manufacture, attracted duty as manufactured tobacco, but tobacco cut at twenty-nine to the inch could be imported as unmanufactured, attracted less duty and was equally suitable for the purpose.[13]

Of even greater concern to the select committee was the apparent manipulation of prices by the British American Tobacco Company (BAT), the parent company of Wills. From the evidence of the Customs Department, it seemed that BAT was charging Wills NZ exorbitant prices for imported tobacco in order to reduce the domestic profit and thereby avoid paying New Zealand tax.[14] The price of bright flue-cured tobacco from the US being imported into New Zealand was stated to be US50 cents per pound. The same leaf was being imported into Australia for US35 cents per pound.[15] After lengthy questioning Smith admitted that BAT was indeed charging Wills NZ more than the market value for imported tobacco but was unwilling to elaborate on this statement. Chairman George Black asked whether Wills in fact wanted to use New Zealand-grown tobacco at all. Smith responded that Wills had been manufacturing and selling products made with the local tobacco for only two years and they could not risk using higher proportions until the market was ready.[16]

This was belied by the success of the products of the National Tobacco Company, which contained significant percentages of domestic leaf, and which had secured a large share of the local market – so much so that Wills had negotiated supplies of National's products to supply its retail outlets following customer preference. Tapawera grower F.W. Gibbs claimed that tobacco retailers sold 'ten tins of National tobacco to every one of Wills'.[17] Wills, of course, concentrated on cigarette manufacture and, while many smokers were still smoking pipes and rolling their own cigarettes, it may have been necessary to use other companies' cut tobacco products to provide a full range in its outlets.

The committee requested financial and commercial information from all concerned

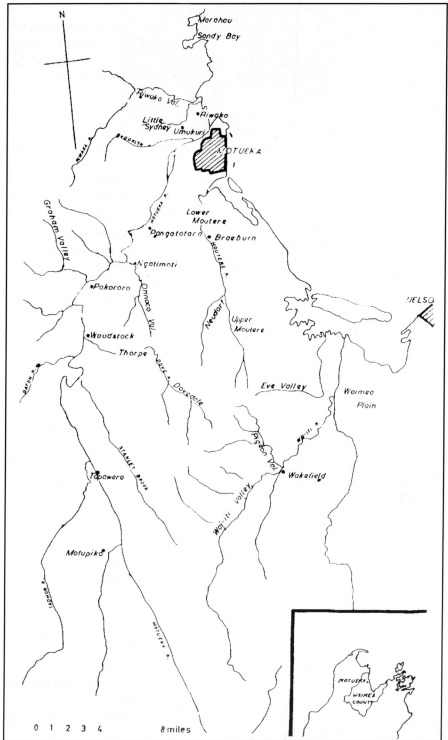

Location map showing major tobacco growing areas in the former Waimea County.

Gerhard Husheer. Regarded by many as the founder of the New Zealand tobacco growing industry.

Cecil C. Nash. Began growing air-dried tobacco at Brightwater, Nelson in 1920. Nash became an influential advisor for the National Tobacco Company.

Charles Lowe. Pioneer tobacco grower in the Motueka area.

A copy of this Prospectus has been filed at the Office of the Deputy Registrar of Companies at Napier, Hawke's Bay, this 25th day of January, 1915.

PROSPECTUS

OF

THE NEW ZEALAND TOBACCO COMPANY LIMITED.

One of the Company's Drying Sheds.

The cover of the prospectus of the N.Z. Tobacco Company Ltd.

PROSPECTUS

OF

THE NEW ZEALAND TOBACCO COMPANY LIMITED.

A Private Company duly incorporated under The Companies Act, but about to be re-registered as a Public Company. (This Prospectus is issued in respect of the said intended Public Company.)

CAPITAL £25,000.

Divided into 25,000 ordinary shares of £1 each.

Shares already issued 9,000.

Shares now offered for public subscription 16,000, to be paid by calls of 2s. 6d. on application, 2s. 6d. on allotment, and the balance by calls of not more than 2s. 6d. at intervals of not less than two months.

DIRECTORS.

R. F. SORENSON, Picture Theatre Proprietor, HASTINGS, Chairman.
F. C. HARTSHORN, Manager Excelsior Dairy Coy., HASTINGS.
W. B. JONES, Manager Booth, McDonald and Coy., Ltd., HASTINGS.
J. COWAN (Le Quesne and Cowan, Plumbers), HASTINGS.
J. T. BLAKE, Licensed Interpreter, HASTINGS.
W. KARAITIANA, Sheepfarmer, HASTINGS.
D. DONOVAN, Settler, HASTINGS.
GEO. EBBETT, Solicitor, HASTINGS.
GERHARD HUSHEER, Tobacco Expert, Hastings.

MANAGER.

GERHARD HUSHEER, HASTINGS.

BANKERS.

THE BANK OF NEW SOUTH WALES, HASTINGS.

AUDITOR.

C. J. TIPPING, HASTINGS.

SOLICITORS.

EBBETT AND BANKS, HASTINGS.

SECRETARY.

F. C. HARTSHORN, HASTINGS.

REGISTERED OFFICE: STATION-STREET, HASTINGS.

List of directors of Husheer's original N.Z. Tobacco Company Ltd. 1912.

An early field of tobacco at Husheer's plantation, Haumoana, Napier, c1913.

Drying sheds on the Haumoana plantation.

Packing the dried leaf at Haumoana.

Crated tobacco arriving at Husheer's Napier factory c1915.

Planting out seedlings, Riverhead c1921.

Hoeing along rows of young plants, Riverhead c1921.

Log kilns near Myrtleford, Victoria similar to those in Carolina which inspired Lowe to consider flue curing in New Zealand.

Charles Lowe harvesting an early crop.

Colourful advertisement for Gold Pouch tobacco.

'Pulling plants' in the seed-beds at Riverhead, Auckland c1921.

Threading harvested plants onto stick ready for drying. Riverhead c1921.

Hanging the loaded sticks in the drying shed. Riverhead c1921.

A group of Nash's early converts at Riwaka c1922.

The horse was an important part of cultivation for many years.

Sun-conscious women at the seed-beds 1920s.

Children played their part, seen here on an early planting machine.

Satisfied Dovedale grower Hugh Thorn stands among his growing crop.

Even the family car was called into service for the harvest.

Women and children were vital in the workforce.

Nelson Tobacco Growers' Association receiving shed in High St. Motueka c1926. Taken over by W.D. and H.O. Wills in 1928.

Brand new fibrolite kiln c1928.

Modern kiln complex of Tauranga Citrus and Tobacco Co. located at Papamoa c1931.

Sticks of tobacco drying in the open air.

Tying leaf onto sticks ready for drying. Note leaves now dried singly.

Carrying sticks to the drying kiln. Rotorua c1929.

Women tying tobacco at the Arawa plantation, Rotorua c1929.

Staff at Riverhead plantation c1933.

Working ground at Riverhead c1933.

companies. Most provided the required documentation, but neither National Tobacco nor Wills was willing to provide full commercial details. Wills, in fact, could not even supply a balance sheet as these were not published in New Zealand – apparently this was not required as there were no New Zealand shareholders in the company.[18] Such information as Wills did provide reinforced the belief that its profits were being manipulated to avoid New Zealand tax. Although operating on a much larger scale than National, Wills paid almost exactly the same amount of income tax in the year under review.[19] Battling with the complexities of international commercial activity on the one hand and the self-interest of individual farmers on the other, the committee produced a set of twenty-two recommendations and suggestions. Their intention was to address most, if not all, of the growers' grievances while allowing companies the commercial leeway they required.

Presenting an impressive report in October 1930, George Black covered the complexities encountered by the committee and put forward proposals designed to stabilise the New Zealand industry. With regard to tariffs, the committee recommended import duty on unmanufactured tobacco should be 3s per pound and on cut tobacco 5s 6d per pound, with excise duty to be abolished on cut tobacco. In order to clarify the imposition of these duties, the committee recommended an amendment to the Tobacco Act of 1908 to clearly define fine-cut tobacco as being tobacco with forty cuts or more to the inch. Likewise, unmanufactured tobacco was to be redefined and was to include leaf that was imported in strips – stripped from the stems.

To foster better management of tobacco growing, the committee proposed that supervision of tobacco culture should be removed from the Department of Agriculture and placed under a tobacco expert to be known as Director of Tobacco Culture, appointed by the government. To assist this director, the committee suggested the formation of a Tobacco Advisory Council consisting of four representatives of the tobacco-growing districts, with the Director of Tobacco Culture as the chairman. It was also suggested that the Advisory Council have the right to invite manufacturers to participate in its meetings, as well as the Comptroller of Customs and the Secretary of the Department of Industries and Commerce. The seeds of the future Tobacco Board were embodied in this plan. Special attention was paid to the question of encouraging and developing an export market for New Zealand tobacco, both in Britain and Australia.

The Te Atatu growers were not forgotten. The committee recommended that they be paid the equivalent of the government-guaranteed prices for export tobacco with no deductions for shipping and related charges. The problem of speculative growing companies was also addressed. Firm recommendations were made that statements made by government advisers should be cleared by heads of departments before they could be used in company promotion. The committee also felt strongly that companies issuing bonds for public subscription should be registered under the Companies Act as well, and went so far as to issue a warning to the public about investing in such companies.

For growers, the benefits would be structural rather than direct. Tariff alterations and the formation of the Advisory Council, together with the newly-created position of Director of Tobacco Culture, were intended to bring material advantage for the growing industry. The council itself was seen as an opportunity for growers to have 'a voice in their industry… and also bring together the [sic] several interests to further a common object'.[20] This hope can be seen as somewhat optimistic as the inquiry highlighted the immense divide between the interests of growers and companies. Growers sought stability and an assured market for their produce. Companies were absorbed with jockeying for market share, protecting commercially sensitive information and lobbying for the lowest possible duties. The 'common object' was far more elusive than even the conscientious and knowledgeable George Black realised. With tobacco being grown under contract to specific buyers, the power remained always with the manufacturer, whose interests were not necessarily favourable to the grower. This situation was to be the backdrop to the entire history of growing tobacco in New Zealand.

Political debates on the industry continued each year. The government, struggling with the effects of the Depression, saw little opportunity to offer growers protection. Opposition members concentrated on the benefits of employment and successful small farmers. H.T. Armstrong, member for Christchurch, toured the Motueka district and believed it was 'the finest example of what can be done by close settlement'.[21] It had been achieved, he said, without government assistance. Keith Holyoake maintained that duties on imported unmanufactured tobacco should be increased as this was the competition for local growers. He cited South Africa and Australia, where duties were structured to protect the domestic industry, while still raising the necessary revenue.[22] But the government was 'hard pressed for revenue' and the decline in cigarette consumption left no option but to increase tax on other tobacco products.[23] Thus, despite the select committee's intensive work and their recommendations for changes to the industry, almost nothing was done at government level to implement these recommendations.

In 1931 growers continued their usual round of seasonal tasks in an atmosphere of uncertainty, knowing that companies were still reconsidering their positions with regard to New Zealand-grown tobacco. Although their numbers and production were rising, many farmers continued to grow other crops such as raspberries, hops and beans to spread their risk of loss or crop failure. New tariffs also played a part in this. During 1931, changes to duties brought import duty on unmanufactured tobacco down from 3s per pound to 2s 6d per pound while the excise on cut and plug tobacco rose from 1s 8d per pound to 3s 8d. This had a double impact on domestic growers, as their product competed with imported tobacco which was often stripped and stemmed, and most New Zealand-grown leaf was used in the manufacture of cut and plug tobacco. The higher excise on imported tobacco was partly offset by the reduction in duty on raw leaf but for domestic leaf, the 2s per pound rise was devastating.

Kiln growers were the hardest hit. As Wills implemented its list of 'approved growers', almost all the Moutere kiln growers disappeared, as did a number of Dovedale growers.[24] Some air-dried growing continued in the Moutere area but by 1932 very little was being grown in this district at all. After the 1930-31 season, however, grower numbers and acreages began to rise again, possibly in reaction to the effects of the Depression on other farm produce and the perception that companies were, in fact, purchasing virtually all the tobacco produced by local growers. The number of bales of tobacco shipped over the Motueka wharf reflect this growth – 4903 bales left for the North Island factories over the local wharf in 1932, a 63 per cent increase on the previous year. For those able to retain contracts, tobacco growing was providing an assured income in a period when wool, meat and butter were oversupplied and fetching low returns. As a result, Motueka prospered when the rest of the country was '… struggling under the most difficult conditions ever experienced in this country'.[25]

The going was still tough for many, however, and the 1932-33 season was very dry well into January. Harvesting began early and by mid-January Cyril Fry had two kilns in and John Duncan had cured one. Strong winds in that month laid much tobacco low, but by the end of January the plants were standing up again. Picking was in full swing, with the usual influx of seasonal workers, although some farmers could have done with more help. The blight known as mosaic, which was to be of increasing concern throughout the 1930s, was spreading rapidly by February and a leaf spot known as 'frog's eye' was also evident. By then, however, the balance of the crop was ripening well, F. Alan Hamilton displaying a 'fine crop of Wills' Special'[26] and National growers harvesting their air-dried leaf.

Also in February two growers from Marlborough, Messrs Rutledge and March, visited the district. They were each growing three acres for Loyals of Auckland, a company managed by E.V. Owen. The Marlborough growers had planned to plant fifty acres, but heavy winds and an infestation of cutworm reduced the area to twenty-five acres. Once again would-be growers in Marlborough were to be disappointed. When two Umukuri growers visited Marlborough in March they found the crops tattered by wind and the strong soil producing dark, rank leaf. The maize windbreaks used by Marlborough growers had not protected their crops from the buffeting and relentless winds experienced by that area.[27]

The big news of early 1933 came in mid-February. Once again excise was to be increased, this time only on leaf used for the manufacture of cut and plug tobacco for use within New Zealand. No corresponding increase was imposed on leaf imported for use in cigarettes, nor were these products subject to any increase in excise or sales tax.[28] Growers were forced to conclude that government measures were intended to suppress the smoking of roll-your-owns and to encourage cigarette smoking, thereby penalising New Zealand growers who were, at this stage, largely producing leaf for cut and plug tobacco. There now began a series of deputations to Wellington to make submissions and pleas for government support for an industry which was of growing importance in the local economy. Chairman of the Growers' Association, a local solicitor named Spencer Smith, was accompanied on this first delegation by the association secretary, Bertie Rowling, as well as two local businessmen, Harry Manoy and James Maxwell. Two growers, James F. Balck of Woodstock and Cyprian Brereton of Orinoco, also joined the delegation. Local members of Parliament Keith Holyoake and Harry Atmore headed the group, which carried the message that the new duties would 'ultimately crush the industry'.[29] Telegrams had been sent to several MPs around the country enlisting their support for the delegation and a telegram of strong support for growers was sent to the Minister of Customs, Gordon Coates, by the mayor of Motueka, Rupert F.L. York.

Meeting Coates and the Prime Minister, George Forbes, the delegation stressed the financial impact of tobacco. Seven hundred growers in the Motueka area were producing tobacco worth £180,000 – of this, £85,000 was paid out in wages, a major consideration in times of alarming levels of unemployment. The government made it very clear that the prime motivation behind the increases was to raise more revenue. In response, the growers' delegation recommended the imposition of an extra eight-pence per pound on imported cigarettes and raw leaf. This would have the double effect of protecting local leaf and raising further revenue. The Prime Minister assured the deputation that their submissions would receive full consideration and the visitors went home hoping for the best.

At a public meeting in Motueka on 20 February 1933 Rupert York fiercely attacked the new tariff measures, referring to the government as 'an octopus'. Although many may have agreed with him, the mayor was criticised by Horatio Everett, who believed it was less than sensible to attack those you hoped would come to your aid.[30] This argument did not impress Keith Holyoake, who attacked the new duties in Parliament on 28 February, but to no avail. A few days later the Minister of Customs stated that the object of the bill was to increase revenue and that the duties would not be detrimental to local growers.[31]

Back on the farm, growers could only shake their heads and carry on with the season's harvest. In March the *Pearl Kaspar* delivered 150 tons of coal for kiln furnaces and in Ngatimoti the heaviest crop from the greatest acreage so far was being harvested. Kiln fires were common – at Ngatimoti George Beatson lost his kiln and at Pangatotara John Canton was also hit. At Umukuri Syd Rowling was a little luckier, the fire in his kiln being discovered in time for the building to be saved. By late March most of the kiln tobacco was harvested and National growers were cutting their plants and building or enlarging drying sheds to cope with the bumper crop.

Throughout this time the local position was being compared with that in Australia and elsewhere. Almost every week articles appeared in the *Motueka Star-Times* regarding conditions in other tobacco-producing countries. Australian growers were struggling with similar problems over tariffs while grower numbers had skyrocketed between 1930 and 1932, from 660 to 4764. Acreage in Queensland alone rose from seventy-one acres in 1930 to 6300 in 1933. With cheap American leaf pouring into the country,[32] Australian growers found themselves priced out of business. New Zealand growers were kept well aware that a similar fate could easily be in store for them.

The vulnerability and relative insignificance of the local industry was further illustrated by a talk given by Buxtons manager Fred O. Hamilton on his return from the Ottawa Imperial Conference in April 1933. He told local growers that the US had two million acres in tobacco cultivation and 400,000 growers, while Canadian growers had planted 55,000 acres. Although he found conditions primitive in the US, with many kilns still being of logs with mud caulking, and thought the quality of

Canadian leaf well below that of New Zealand tobacco, Hamilton pointed out the long history of expertise and market domination of the American growers. He warned his audience of their precarious position in light of the discussions at the Imperial Conference.[33] The conference addressed tariffs and protectionist policies and, as in more recent years, New Zealand was warned to resist imposing tariffs intended mainly to protect domestic industries.

No favourable response having been received from the government by the time the 1932-33 crop had been graded and dispatched, growers again decided to send a delegation to Wellington. This time they planned to present submissions to the Tariff Commission, which was examining the tariff structure of many industries, and to appeal again to the Prime Minister. Spencer Smith and Keith Holyoake provided the commission with the history of tariffs, both import duties and excise tax. They demonstrated that whereas the import duty had fallen dramatically since 1930, excise tax, which affected local tobacco, had risen by 132 per cent in the same period and now stood at 4s 01/2d per pound.[34] Growers were again suggesting an increase in the duty on imported unmanufactured tobacco to enable New Zealand leaf to compete economically.

The chairman of the commission, Dr Craig (who was also, coincidentally, Comptroller of Customs), had also appeared before the 1930 select committee where he had given evidence that the local growers were being subsidised to the tune of £150 each through favourable tariffs;[35] his estimate had now risen to £192 per grower. He maintained that domestic growers were attempting to force the New Zealand consumer to smoke their tobacco. The delegation argued that if duties had to be levied for revenue purposes, then at least they could be arranged to foster the local industry rather than American growers.[36] Keith Holyoake submitted that 1000 more families could be placed in the tobacco-growing industry if conditions were made more favourable.

The government was indeed in a somewhat peculiar position. On the one hand, Nelson growers were being instructed to face the facts of international competition and the government's inability to provide support and protection. On the other, the Departments of Agriculture and Labour, along with the Unemployment Board and the Small Farms Scheme, were heavily involved in a scheme in the Bay of Plenty. In the Pongakawa Valley, inland from Te Puke, efforts were being put into settling unemployed families to produce tobacco under the guidance of Charles Lowe. As Nelson growers struggled with officialdom's lack of response to their pleas for support, the government was virtually subsidising tobacco production in Pongakawa where the hapless settlers had their own troubles to occupy them.

NOTES
1. Figures taken from growers' production returns furnished to the 1930 Committee of Enquiry.
2. List of approved growers, 22.7.30, National Archives file LE 1 1930/16, History of the Industry.
3. All information on contract cuts from Minutes of Evidence, Committee of Inquiry into Tobacco Growing, 1930, pp.573, 479, 261, 265-67.
4. Letter from Wills to Buxtons, 22.7.30, National Archives file LE 1 1930/16, History of the Industry.

5. Minutes of Evidence, Committee of Inquiry into Tobacco Growing, 1930, p.234.
6. Ibid, p.140.
7. Ibid, p.98.
8. Ibid, B18.
9. Ibid, p.253.
10. National Archives file LE 1 1930/16, History of the Industry, pp.16-17.
11. Minutes of Evidence, Committee of Inquiry into Tobacco Growing, 1930, p.552.
12. Ibid, p.586.
13. Ibid, p.42.
14. National Archives file LE 1 1930/16, History of the Industry, p.8.
15. Ibid, p.10. In the parliamentary debate on the committee's report, one MP, Mason, described this as a 'simple trick' to defraud the tax system. Parliamentary Debates, 1930, p.1004.
16. Minutes of Evidence, Committee of Inquiry into Tobacco Growing, 1930, pp.563, 565.
17. Correspondence, Committee of Inquiry into Tobacco Growing, 1930, F.W. Gibbs.
18. Minutes of Evidence, Committee of Inquiry into Tobacco Growing, 1930, p.562.
19. Minutes of Evidence, Committee of Inquiry into Tobacco Growing, 1930, pp.6, 10.
20. Ibid, p.24.
21. Parliamentary Debates Vol. 242, p.156.
22. Ibid, Vol. 235, pp.740-41.
23. Ibid, p.713, Coates.
24. Waugh, J.R., The Changing Distribution of Tobacco Growing in Waimea County, MA thesis in geography, 1962, Figs 26 and 27, p.42.
25. Motueka Star-Times, 26.7.38.
26. Motueka Star-Times, 10.1.33, 13.1.33, 24.1.33, 27.1.33, 10.2.33.
27. Motueka Star-Times, 7.2.33, 14.3.33.
28. Ibid, 14.2.33.
29. Op cit,
30. Ibid, 24.2.33.
31. Ibid, 3.3.33.
32. Motueka Star-Times, 23.12.32, 28.7.33, 8.9.33, 5.1.34.
33. Ibid, 28.4.33.
34. Ibid, 14.7.33.
35. Minutes of Evidence, Committee of Inquiry into Tobacco Growing, 1930, p.1.
36. Motueka Star-Times, 14.7.33.

CHAPTER SEVEN

Trials and Tribulations in the Pongakawa Valley

As unemployment soared during the Depression, the government was forced to look at any means it could to place people in some type of productive occupation. The concept of growing tobacco for export had been part of the Department of Agriculture's programme since the establishment of the Motueka Growers' Association in 1926, along with the government guarantee scheme which underwrote the Motueka initiative. Charles Lowe's enthusiasm for his duties as government instructor had taken him up and down between Northland and Rotorua in the late 1920s. During this time he had taken note of various localities that might be suitable for tobacco growing.

Among these was the Pongakawa Valley in the Western Bay of Plenty where the Crown owned a significant tract of land which was, at that time, largely unsettled. From the 1890s European settlers had attempted pastoral farming with little success. Trace-element deficiency caused bush sickness in stock and many settlers simply abandoned their properties. Much of the valley reverted to fern and manuka, with ragwort causing concern to the few remaining farmers. By the early 1930s, when Charles Lowe visited the area, it hardly looked promising for any agricultural project. Nonetheless, planning proceeded for a tobacco plantation worked by unemployed families relocated from other areas. Under the Small Farms Scheme, the land was subdivided into forty small farms of five acres each in 1932. The first two basic cottages were completed by October 1932 and the plots were balloted for. Collaboration between the Departments of Agriculture and Labour resulted in the selection of forty families and by the end of October, three families were in residence and planting had begun.[1]

These first families must have felt like pioneers as they trekked into the isolated valley. The houses the settlers occupied were tiny two-bedroom cottages, built under contract 'at slump, cut-throat prices',[2] for which each family paid five shillings a week in rent. The valley itself was hardly a rural idyll with its scrub-covered slopes and raw earth floor newly worked-up ready for tobacco planting. There were none of the amenities familiar to the erstwhile townsfolk – no shops, no transport, no telephone and no school. The potholed pumice road rarely saw a car or truck.[3] Later, one of the tobacco settlers, Percy Maude, whose 1923 Dodge was one of the few cars in the valley, taxied many women to the Te Puke maternity hospital for the birth of their babies. The new settlers obtained milk from a dairy farm in the centre of the valley – delivery was by bicycle, with the milk being doled from a cream can into billies with a dipper.

The diverse backgrounds of the settlers hardly fitted them for the task ahead. An artist, a former garage owner and a support worker on an Antarctic biological mission were amongst the first would-be tobacco farmers of Pongakawa, while Dorothy Maude was a concert pianist.[4]

Needless to say, many did not stay long, but those who left were replaced by others hopeful of a better life than they were experiencing as urban unemployed. They were to be sorely disappointed, as the conditions improved only slightly over time and the returns were minimal. The male farmers remained on the equivalent of relief wages (37s 6d per week) until 1935, and those who arrived before the first twenty cottages were completed were housed in tents. The water supply was still causing complaint in 1936 and the lack of experience and specific skills caused some disagreement between farmers.[5]

Despite the drawbacks, the early tobacco crops were relatively successful and at the end of 1932 Charles Lowe described the plants as 'doing well'.[6] With flue barns and drying sheds constructed, the first season yielded 15,000 pounds of leaf.[7] The settlers, too, began to improve their surroundings. A school was set up late in 1933 – children climbed into the loft of one of the tobacco kilns for their lessons. The Pongakawa settlers were most unhappy that many of their children had been without schooling for months. Many felt that their youngsters were running wild and would be severely disadvantaged, especially those nearing their Proficiency exams. After intense pressure on Department of Education officials from the residents, a new school building was erected and opened in April 1934. The disparate group of novice farmers and their families was becoming a community in itself. The valley was still isolated and school children found the rare passing of a car, with its accompanying cloud of dust, an intriguing distraction from their work. At the peak of tobacco production in the little community, in 1936, there were thirty children attending Rotoehu School.[8]

But the peak soon passed. The area under cultivation fell by almost half between the 1936-37 and the 1937-38 seasons. The isolation, with its attendant privations and difficulties, compounded with improving conditions in the general employment market, caused many of those who were reluctant agriculturalists to move out of the valley, back to more familiar surroundings. Tobacco growing continued in the valley but the settlers' complaints were becoming continuous and insistent. The 1937-38 season saw the government handing supervision of the project over to the Forestry Department, evidently believing that tobacco cultivation and the new forestry nursery could be run jointly. Charles Lowe was by then back in Motueka supervising the establishment of a reconditioning plant, and the emphasis on export tobacco had once more shifted to that area. During the 1938-39 season the remaining Pongakawa tobacco farmers received two blows that virtually ended the experiment. In December 1938 windstorms swept down the narrow, funnel-shaped valley creating massive dust clouds from the eight kilometres of cultivated land. Less than a month later a gale, followed by 125 millimetres of rain in four hours, hit the embattled valley. It was clear there was no tobacco gold to be found in the Pongakawa Valley, and the remaining settlers were given the option of staying on to work in forestry development and plantation or repatriation to their urban origin. While a number chose to leave, many elected to stay and formed the nucleus of the forestry settlement.

The government's involvement in this scheme through various departments and boards was a mixture of misguided and often inept entrepreneurialism and pragmatic, but unrealistic, attempts to deal with unemployment. Charles Lowe, as government tobacco instructor, was convinced that the light pumice soils would produce satisfactory tobacco and even believed that the area was too small. Projecting (in 1932) that the area under cultivation could reach 570 acres by 1936-37, he also identified nearby valleys of Pikowai and Kaikokopu as suitable for development. Basing his calculations on Brodie's forecast for Motueka, which had been exceeded within five years, he suggested that up to 110 families could be placed in the area. Lowe thought that the Hopuhopu Military Camp could be rented to provide facilities for grading, processing, packing and preparing the tobacco for export.[9] This, of course, never came about.

The tobacco produced at Pongakawa was to be prepared for export by Loyals Tobacco Company, of Auckland. This association was to prove costly and slightly embarrassing for the Department of Agriculture and, despite his avowed impartiality, Charles Lowe became embroiled in several difficult situations involving the company. Loyals was registered in 1930 with a subscribed capital of £30,000 and with the aim of processing local tobacco for export. In 1933, J.A. Campbell, Director of Horticulture, wrote to Lowe to the effect that Loyals was in trouble.[10] Later that year a Treasury loan was arranged for Loyals to cover the difficulties it had encountered as a result of mismanagement of the payment of excise duty.[11] The maintenance and support of Loyals was crucial to the government scheme at Pongakawa as, according to Lowe, the company purchased all the leaf from the 1934 and 1935 harvests and, what is more, paid for it.[12] The existence of the Treasury loan suggests, however, that the government not only financed the settlement and the production of the leaf, but also underwrote Loyals' purchasing power. Government financial losses over the duration of this scheme were inevitably a major factor in discontinuing the project.

Pongakawa was not the only area in the Bay of Plenty to be involved in tobacco cultivation. Further north, towards Tauranga, the United Tobacco Company (Tauranga) Ltd (formerly Tauranga Citrus and Tobacco Company) had begun its planting in 1929-30 with seven and a half acres, which produced 6000 pounds of leaf. Under manager Billy Owen the company increased its plantings to fifty acres and erected a modern, five-kiln, curing barn complex. Production from this plantation was intended to be used both on the local market and for export to Australia. As with other companies formed to grow tobacco on a large scale, United failed to secure a market for its leaf and although it initially produced some of the best North Island leaf,[13] United's management practices evidently slipped. Wills' adviser stated that 1932-33 crop at Papamoa 'never did ripen' and that United had spent £6000 and produced nothing.[14] In 1934 Lowe observed that the latest crop was in a poor state. The fifty-six acres under cultivation was stunted in appearance from lack of manure at planting and there had been no cultivation since. Feeling that the ground was too cold, Lowe considered that planting should have been done in ridges,[15] a practice that did not come into general usage in the Motueka area until the 1960s. With 10,000 pounds of trash tobacco on hand, things got no better for United and the company disappeared from view after the 1933-34 season.

Wills' own trials in the area were also unsatisfactory. Hoping to produce 30,000 pounds from a fifteen-acre trial, the company sought farmers in three different areas. Up to five acres would be grown in each area and flue-cured in kilns financed by Wills. Farmers were initially guaranteed £50 per acre, giving a profit of £10 per acre, but the guarantee was later reduced to £40. The company was hardly enthusiastic about the project – the manager, R.B. Smith, was strongly against starting any scheme without good prospects of success. 'Everything in the history of tobacco growing in the North Island up to date is a warning against any further growing unless all the conditions are as favourable as possible.'[16] Conditions were not at all favourable, all tobacco in the Bay of Plenty being damaged by wind and rain in 1932-33. Wills' overall summary of its trials in this area described the Bay of Plenty as cattle-sick country, probably with some sort of soil deficiency.[17] Compared with Nelson, the weather was too hot and humid and the nights too warm for the production of good-quality tobacco.

The Bay of Plenty experiences, including those of the Arawa at Rotorua, also illustrated the difficulties of growers without an assured contract or market for their tobacco. The concentration on production for export was another vital factor in the demise of these efforts. It was obvious to others in the industry that attempts to produce export tobacco in New Zealand under prevailing conditions were doomed to failure. Although the quality of domestic tobacco was open to various opinions, the main factors against a viable export industry were the internal cost of production and the freight charges involved in transporting the crop to the international markets. Even tobacco produced through schemes such as Pongakawa, where the wages were as low as any in New Zealand, could not match low prices of crops from countries using unpaid or lowly paid workers. The Empire Tobacco Corporation claimed that its proposed large-scale operation would perform far more efficiently and cost-effectively than the individual small farmers in other areas[18] but there was little chance of producing for a sufficiently low cost to compete internationally. As the New Zealand economy improved, costs were likely to rise rather than decrease, further disadvantaging the crop in relation to low-cost producers. The 1930s were to see the last real attempts to cultivate tobacco commercially outside the Motueka-Nelson district and the history of the crop from this point centres almost entirely on this region.

NOTES
1. Personal papers of Charles Lowe.
2. A History of Rotoehu Forest, from material supplied by June Drabble, Pongakawa.
3. Potter, P., A History of Rotoehu School, p.5, supplied by June Drabble.
4. Interview with Renee Benner, Pongakawa Valley, 30.9.95.
5. Personal papers of Charles Lowe. Given the inexperience of most of the settlers it seems that those with even a modicum of previous work with plants of any kind felt entitled to positions of authority.
6. Ibid, letter to J.A. Campbell, Director of Horticulture, reporting on progress, 31.12.32.
7. From papers supplied by June Drabble.
8. Potter, op cit, pp.7, 12.

9. Personal papers of Charles Lowe, 6.10.32.
10. Ibid, letter from Campbell, 30.10.33.
11. Lowe explains this in a letter to Campbell in 1935. Loyals had used 'trashy' tobacco to make immediate sales. Some of this was unsuitable and was returned and further mixed with new tobacco. Having paid duty on releasing the leaf the first time, Loyals did not think to pay again on re-releasing. After 18 months the company received a writ for £45,000 penalty excise duty. This was reduced to £9500 and again to £7800. Ibid, 3.9.35.
12. Ibid letter to Campbell, 16.6.36.
13. Ibid, 4.4.32.
14. Letter from Gracie to R.B. Smith, 11.2.33, Motueka Museum.
15. Personal papers of Charles Lowe, 8.1.34.
16. Personal papers of Charles Lowe, letter from R.B. Smith to Gracie, 27.5.32.
17. Personal papers of Charles Lowe, letter from Gracie to Smith, 11.2.33.
18. Brochure published by the Empire Tobacco Corporation, c. 1930.

CHAPTER EIGHT

A False Sense of Security?

Back in Motueka, with the 1932-33 crop graded and sold, growers again turned their attention to the political situation of their industry. One hundred and sixty growers attended the annual general meeting of the Motueka District Tobacco Growers' Association, held in Motueka on 15 September 1933, where they were reminded that the main factor they had to contend with was the cost of production in the US. Since American growers were 'even subsidised by the government', local cost of production should be kept under constant review. Some growers were said to be 'imbued with a false sense of security'.[1] For some years they had been receiving prices for their leaf which, while providing a fairly good level of profit, were totally dependent on manufacturing companies. To achieve a standard price for domestic tobacco, growers needed to create and maintain unity amongst themselves and persist in attempts to secure protection against foreign imports.

The association was already operating a ward system of representation, which was to come into force under the future Tobacco Growers' Federation. Growers from each district had two representatives of their own selection on the executive of the association. The wards, as then constituted, were:

> Number One Ward: Waimea-Tapawera
> Number Two Ward: Stanley Brook, Woodstock and Dovedale
> Number Three Ward: Pokororo, Orinoco and Pangatotara
> Number Four Ward: Motueka and Riwaka
> Number Five Ward: Lower and Upper Moutere

Under the federation these five wards would be reduced to four and geographically rearranged.

The question of unity among growers would underlie much of the history of both the association and later the federation. In an organisation representing the interests of up to 700 individual farmers, there were bound to be opposing views and a variety of interpersonal relationships. Within the relatively small area of the Nelson province there were, nevertheless, regional differences and rivalries. As in many other organisations, those growers from outlying areas often felt that they were not told certain things and that the association was controlled by 'townies'.

In 1933 the issue of unity revolved around claims made by Cecil Nash that the executive was dominated by the three businessmen, Spencer Smith, James Maxwell and Harry Manoy, who had been prominent in delegations to Wellington. Nash, who was a member of the executive, had addressed meetings of growers around the district to the effect that the delegation's submissions for the imposition of eightpence per pound import duty on raw leaf had not been agreed to by growers and were not in the best interests of the growing industry. It was claimed that he had attended at least one of these meetings with a wire from the National Tobacco Company to this effect in his pocket, and that the company had referred to the growers as 'mad' in asking for the extra duty.[2] In effect, Nash alleged that the delegation had altered the focus of the submissions previously agreed to by the full executive, and the new stance was not one which the companies would support. Added duty of any kind, of course, was not in the best interests of manufacturers, whether or not it benefited growers.

The allegations were strenuously denied by those concerned – Spencer Smith declared that 'being on the executive [was] no sinecure'. A great deal of time and energy was devoted to association matters for which members 'apparently receive[d] little consideration'. As for the charge that businessmen were running the executive, Smith was applauded by his listeners for his assertion that: 'Tobacco is the backbone of the district, and we are doing our best in the interests of the growers.'[3] The AGM appeared relatively satisfied for the 'businessmen' to remain in their positions. All were re-elected, although at an executive meeting following the general meeting, two growers' members, B.T. (Bertie) Rowling and Alfred E. Fry, were added to the emergency committee which had previously consisted of the three Motueka businessmen.

That little storm over, the growers turned their attention to their local member of Parliament, Keith Jacka Holyoake, who gave a lengthy explanation of his actions regarding the legislation introducing new duties and excise tax. As the bill had been introduced by his own party, Holyoake had been placed between government policy and his constituents' interests. The young MP had succeeded in moving an amendment to the proposed bill and in dividing the House on the question. Holyoake made the government's motivation very clear – the increase in the use of cut and plug tobacco through the Depression, at the expense of cigarettes which attracted higher duty, had severely affected revenue collection. The new duties, which primarily applied to raw tobacco, were designed to address this situation and for this reason cigarettes had not attracted a similar increase. In answer to a grower's question as to whether he had followed the Labour Party in this matter, Holyoake replied, to applause, 'I moved the amendment. I did not follow the Labour Party – I made them follow me.'[4] The amendment had been lost, the government legislation progressed, but Keith Holyoake retained the confidence of his electorate.

Growers, however, were no better off. With the disastrous prices received for air-dried tobacco in the previous season, some were pressing for growers to be present when tobacco was assessed by companies. With restricted local buying, it was left to companies to be the sole arbiters of quality and price. The process was labelled 'evil' by one newspaper correspondent, who maintained that local buying facilities were needed, if only to avoid the expense of shipping worthless leaf 'to the North Island for the construction of its own funeral pyre'.[5]

In the meantime, growers and their families were again into the seasonal round of tobacco culture. Local glasshouses were releasing hundreds of boxes of seedlings, about a week earlier than the previous year. The number of tractors in use by tobacco growers was increasing and there was much discussion on the cost efficiency of various kiln fuels. It appeared that there were wide variations in the cost of curing a kiln of tobacco – while one grower claimed he could cure a kiln for 15s, another listed his per-kiln cost as £5 5s.[6] In Umukuri, W. Evans was making good use of an irrigation system. He was probably one of the first to do so, although some raspberry growers had adapted their existing irrigation systems for use in tobacco fields.[7] Companies were taking a firm line, with Wills announcing that it intended to keep growers to their contract amounts. This was seen by some as a good thing as growers should avoid the dangers of overproduction and the open market, areas where many had been burned before.[8] With grower numbers rising, the possibility of non-contract overproduction was a very real danger.

In November 1933 the Growers' Association was again planning to send a delegation to Wellington, this time to appear before the Customs Tariff Commission, which was considering all duties. By now, however, there were no funds left to finance the deputation and the executive appealed to growers to contribute 2s 6d each to set up a 'fighting fund'. One grower objected to this proposal, saying that, rather than growers being expected to pay out even more for the purpose, the local MP should represent the growers' interests in Wellington.[9] Also in November, National Tobacco Company growers received a circular from the company stating that the company would henceforth accept only long, fair-sized leaf of bright, clean colour. It would also be reducing the acreage offered under contract. In its now familiar style, the company added that it had previously accepted poor-quality leaf in order to assist growers, but that overproduction and growing without contracts had produced a stockpile of leaf, much of which was defective and below the required standard. This factor and the impact of duties and sales tax on product sales had forced it to retrench. In December the National Tobacco Company balance sheet showed that the company was indeed holding substantial stocks of leaf and that acreages would be cut back by half. The dividend paid out to shareholders, however, was a handsome 15 per cent.[10]

A meeting of eighty growers on 30 November endorsed the delegation to Wellington, voting to increase the numbers to five. While three were considered sufficient to appear before the Tariff Commission, it was felt that a larger deputation was advisable in their approach to the government. The selection of Spencer Smith, Alfred Fry and Bertie Rowling was approved by grower D.R. Park, who thought the oratorical strengths of these three were vital to the mission. 'They are a pretty hard mob over there and will tie you up – you want a lawyer to get out of it,'[11] he told the meeting, which erupted in appreciative laughter. The half-crown levy (2s 6d) had so far brought contributions from eighty of the 500-600 growers and more would be needed to send the larger delegation. The idea of a membership fee was now being raised regularly by Dick Selby-Bennetts of Dovedale, and James Balck was now proposing the registration of all growers. Clearly a more formalised organisation was only a matter of time.

Against the backdrop of all this political activity, growers were carrying on with their practical activities. A few, in fact, were already experiencing some of the usual seasonal disasters. A December hailstorm at Pangatotara wiped out forty acres of tobacco, with

Cyril Heaps and Hugh Williams amongst the worst affected, while others suffered damage to a lesser degree. A severe frost a month later destroyed further crops in the Pangatotara area, as well as in Umukuri and parts of Riwaka.[12] Added to the cut-backs imposed by the companies, these natural events shrank the available returns even further.

Despite being in the midst of a busy time on their farms, over sixty growers gathered to hear the deputation report on their meetings with the Minister of Customs and the Tariff Commission. Chairman Spencer Smith, told the meeting that the delegation's conversation with Customs Minister Gordon Coates had been 'quite interesting'. The Minister now appeared far more aware of conditions affecting the growing industry than he had previously demonstrated; Smith thought this was largely due to the 'persistent advocacy' of their local MP, Holyoake. James Balck felt that the Minister was now 'distinctly sympathetic' towards the industry. Coates had asked numerous questions on the effects of higher duties and imports of unmanufactured leaf, and had promised to enquire fully into the matters raised by the delegation. The delegation remained silent on its meeting with the Tariff Commission as they had been 'put on their honour for the time being'. Dovedale grower W. Brislane was unimpressed with this, saying growers might as well 'chuck their money in the sea' as support deputations who were sworn to secrecy. The commission's recommendations would be debated in the House and Smith urged growers to plan to send a large representation to Wellington to support their case.

Following the delegation's report, a long discussion took place on the advisability of setting up a growers' association with the primary object of protecting the interests of the industry. Although there were some reservations, the meeting was generally in favour of moves to establish an organisation which would include all growers and therefore be a stronger vehicle to advance their interests. The influence of companies was plainly evident, however. Alfred Fry was pleased to see that some National Company growers were supporting the proposal, although they were 'supposed to be trembling in their shoes'. Clearly, some were quivering slightly – the motion to form an association was tempered by an amendment to the effect that the companies should be asked for their opinions before further action was taken. Supporting the amendment, J.F. McGlashen commented that National growers would be 'very unwise' to support the move without discussion with their company, 'which had made a definite pronouncement on the subject'.[13] After heated debate the amendment was carried but it would be almost four years before a united growers' association would come into being.

Practicalities still concerned growers far more than political goings-on. By late January 1934, the effects of hail and frost damage were found to be far worse than first thought. Now believed to have destroyed virtually 40 per cent of the total crop, the damage was estimated to be worth £75,000. From Pangatotara to Balck Bridge, all crops except H. Stringer's were practically wiped out. In Umukuri, Joseph Smith had lost nineteen acres and numerous others suffered significant loss. While Dovedale was not hit quite so badly, one farmer, who had been settled under the Small Farms Scheme, had lost all his tobacco and a crop of beans as well. Keith Holyoake was again quick to represent growers' interests, sending telegrams to the Prime Minister and to the Ministers of Agriculture and Employment. Early in February 1934, the Minister of Employment, Adam Hamilton (later to be the first leader of the National Party), visited Motueka, touring several farms to see for himself the effects of the damage. Growers were hopeful that the situation could be met through government-sponsored loans with lenient repayment provisions. Minister Hamilton assured them, 'you can look to the government to do the right thing'.[14] Nothing definite was heard from official quarters until May of the same year.

Meanwhile growers coped as best they could with the damaged plants. In Umukuri, some stripped the frosted leaf off and hoped the tips would size up. Both National and Wills growers were cutting plants off to the first lateral and hoping for a long season with no more early frosts. At Riwaka, the damage proved greater than first thought and it seemed that many would have difficulty filling their contract quotas. By late February, harvesting was proceeding, with a considerable number of kilns already in. Despite the losses, the colour was said to be the best ever and the cut-down plants were progressing well. With the season shortened, workers began to move to other employment – the Brooklyn hydro-electric scheme absorbed so many workers that some Riwaka farms were left short of labour.

The whole disastrous season floundered on until early frosts in March spoiled still more tobacco. All remaining leaf up the 'Shaggery', on the west bank of the Motueka River, was lost, as were the plants which had been cut to encourage lateral growth. Wills' representatives Ramsay and Stott, accompanied by field adviser Ian Hamilton, inspected the crop and confirmed that the company would take only the contract amount.[15] The season of misfortune ground to a close.

A small number of growers had, from time to time, bypassed the problem of contracts and overproduction by making up their own tobacco mixtures and selling their 'home brew' wherever they could. One grower was well known to have taken loads of his brew to the West Coast for sale to men in the relief work camps there. Packaged in paper bags of a pound or so, the illegal tobacco was sold for 7s 6d a packet to the eager buyers. Customs raids did occur, but the cleverer 'smugglers' soon devised a variety of ways of avoiding detection. Successful hiding places included beehives and petrol drums. One truckload stopped and searched on the Hope Saddle kept its secret well – the illicit cargo was hidden in the truck's tyres.[16] Court cases into the 1940s regularly featured the same grower.[17] Obviously the home-brew market did not disappear with the end of the Depression. A trial in Christchurch in 1936 saw two defendants convicted of selling four pounds of illicit tobacco from the Nelson area – the judge deplored the practice of using illegal tobacco to 'water down' commercial products for resale.[18] Many growers made home-brew tobacco to their own recipe for their own use, each with their own blend of ingredients or additives. Honey, wine and tonka beans were among the favourites,[19] although one would-be home manufacturer claimed that 'you could choose virtually anything from the shelves of a chemist shop and put it in'.[20] At a meeting of growers in 1933, Spencer Smith claimed there were dozens in the room smoking their own 'home brew'. He added that, since they all appeared satisfied with their product, there seemed no reason why more New Zealand tobacco could not be used by the companies[21] which were far more skilled at mixing and blending.

The despair of many growers during the 1933-34 season was compounded by the government's announcement, in May 1934, that those who had suffered severe losses in the January frost were to be referred to the department to which they were mortgaged, where applicable. For most this was either the State Advances, or the Lands and Survey Department for those settled under the Discharged Soldiers Act. Sixteen growers were to be referred to the Unemployment Board under the provisions for settlers employed on their own farms and one was to be dealt with by the Small Farms Board. Two growers were not thought to need any government assistance at all. Neither the growers nor their MP were impressed with these proposals for 'assistance' and Keith Holyoake expressed his disappointment to the Prime Minister. For growers it was further confirmation that the struggle for recognition and support had barely begun.

To believe, however, that the tobacco-growing families were sunk in constant and daily depression in the early 1930s would be to misunderstand the nature of rural communities. For women, life on tobacco farms certainly included the long, hard slog of the seasonal requirements of growing tobacco – hours spent crouched while pricking out the tiny seedlings into tobacco beds and more of the same when the plants were ready for 'pulling' and planting. Back-breaking days of planting and replanting, then hoeing, weeding, lateralling and topping were followed by the hectic days of passing, tying and loading kilns. After the harvest was over, winter days, often achingly cold after heavy frosts, were spent grading and hanking ready for buying. Like many other women, tobacco women fitted childbirth, raising a family, and housework around their farm duties. Together with rural women in other parts of New Zealand, tobacco wives somehow also managed to form and run local branches of the Plunket Society, the Women's Division of the NZ Farmers' Union (later of Federated Farmers) and the Country Women's Institute. In Dovedale, Ruth Hall found time to be president of the local branch of the WDFU – similarly in Riwaka, Motueka and the outlying valleys, women fitted community participation around their tobacco work. Meetings were often suspended for the harvest season or held at night to enable more of the busy women to attend. As well, many women were active within their local churches.

For men, the hard work and seasonal disappointments did not prevent them from involvement in sport and community activities. Cricket, athletics, rugby and rifle-shooting were among the sporting pursuits enjoyed by farmers from Marahau to Tapawera. At Riwaka, tobacco family names appear in reports of popular 'community sings'. A. (Fred) Hamilton (a son of F.O. Hamilton), besides his many community positions, also regularly entertained with his impression of 'Rastus' and as a talented master of ceremonies. Dances, fundraisers and RSA activities were all made possible by voluntary effort, much of which came from tobacco growers and their families. In Dovedale, community events were enhanced by Lance Hall and Fred Bennetts, who were a tidy pair at decorating the local hall. Fen Burnett was a regular master of ceremonies at social events, as was Sam O'Hara over the Pigeon Valley hill in Wakefield.

Some social activities revolved around tobacco farming itself. Inter-farm cricket matches were common, as were combined picnics and sports meetings. Rivalry on the sports field was reflected in farm activities. The news that a certain gang had completed a kiln in a particularly short time caused much comment, and competition ensued to 'beat the buggers'.[22] Later, more formal competitions were held, mainly for tying tobacco, as part of local harvest festival celebrations. There were many ways to lighten the onerous nature of the work – smokos provided a brief respite, but end-of-season festivities were the highlight. At Marahau, workers enjoyed many a sing-song at Bloomfield's. Both farmers and workers 'let their hair down', relieved that the job was over for another year.

At this time, too, the district was only minimally served by electricity. In June 1934 it was estimated that more than a hundred prospective consumers would be found 'up the valley' and that tobacco growers would be more than agreeable to 'taking the juice' for drying their tobacco.[23] Such eagerness can be easily understood, given the time and effort spent in fuel collection and kiln watching. With the specific pitfalls of their industry and the unreliability of their market, it is little wonder that the tobacco growers of the Motueka district eventually marshalled their forces late in 1934 when the new customs duties finally came into force.

NOTES

[1] *Motueka Star-Times*, 15.9.33.
[2] Ibid, 15.9.33.
[3] Ibid.
[4] Ibid.
[5] Ibid, 6.10.33, p.3.
[6] Ibid, 29.9.33, 17.10.33, 10.11.33.
[7] Interview with Pat Beatson, 3.8.95.
[8] *Motueka Star-Times*, 21.11.33.
[9] Ibid, 21.11.33, 24.11.33.
[10] Ibid, 5.12.33.
[11] Ibid, 1.12.33.
[12] Ibid, 20.12.33, 23.1.34.
[13] All quotes and information regarding this meeting from *Motueka Star-Times*, 22.12.33.
[14] Ibid, 26.1.34, 2.2.34.
[15] Ibid, 6.2.34, 27.2.34, 20.3.34.
[16] Interview with Dudley Eggers, 21.3.95.
[17] Tobacco Board Correspondence, 27.7.44 and 18.8.48, National Archives, file Tobacco Board 1/1.
[18] *Nelson Evening Mail*, 12.2.36.
[19] Ibid, 22.10.88. Interview with pioneer grower Lance Mytton.
[20] Interview with former grower Kelvin Mytton, Motueka, 4.4.95.
[21] *Motueka Star-Times*, 15.9.33.
[22] Comment from former grower B.Y. (Betty) Fry, Riwaka, 5.12.96.
[23] *Motueka Star-Times*, 22.6.34.

CHAPTER NINE

'The Fight of Their Lives'

By mid-July 1934 the Tariff Commission had forwarded its recommendations to the government. The commission had concluded, somewhat inexplicably, that since tobacco would continue to be imported into New Zealand there was little scope for the local industry to expand. In a statement revealing the true nature of duties on tobacco, the commission further announced that a large amount of revenue had been lost through the popularity of roll-your-own cigarettes. As higher duties on this type of tobacco had been ineffective in raising extra revenue, it was necessary to lower the duties on cigarettes and cigarette tobacco. Excise duty would be raised even higher on tobacco for pipe smokers and hand-rolled cigarettes, as well as on cigarette papers. In Parliament Keith Holyoake railed against the measures, saying his greatest difficulty would be in 'preventing every grower in the Motueka district from coming to Wellington to make his [sic] voice heard against the Bill'.[1] In the tobacco-growing community there were suspicions of undue influence on the commission. A. Kenyon claimed at a Waimea County Council meeting that there must be 'a nigger in the woodpile somewhere'.[2] County chairman J. Corder agreed with him.

The tobacco growers' fight now began in earnest. A meeting in the Majestic Theatre on Friday, 13 July attracted 400 people, including the mayors of Motueka (Rupert York) and Nelson (W.J. Moffatt) as well as representatives of the Nelson Provincial Progress League. The tone of the meeting was forthright and determined, the scene having been set by a stirring telegram from Keith Holyoake and an advertisement for the meeting urging growers to 'Roll up to a man' as their industry was at stake.[3] The association chairman, Spencer Smith, opened the meeting by telling those present that the object was to discuss the position in which the new tariffs would place the industry. The further lowering of duty on imported leaf was an encouragement to manufacturers to use foreign tobacco, and the local growing industry would be 'annihilated'. In his opinion, the tariff changes were intended 'to force smokers off pipe tobaccos and encourage them to smoke made-up cigarettes'. The increased duty on tissue papers used by roll-your-own smokers reinforced this view. The Tariff Commission had, in fact, referred to the manufacture of cigarette papers as an inappropriate industry for New Zealand. No reason was given for this view.

In a long address to the meeting, in which he clearly explained the situation, Keith Holyoake made the point several times that senior government ministers were expecting, if not awaiting, strong representations from tobacco growers. He repeatedly advised them to send as strong a delegation to Wellington as possible. Holyoake exhorted growers to face the facts of their previous history in an industry based on an 'artificial foundation', in that domestic growers had never been compelled to compete on an international basis. The young MP was at great pains to convince his listeners that the new duties had been formulated by an independent commission and were not the product of government policy. A 'belligerent or antagonistic attitude', he declared, 'will not help us one iota at the present time, as the situation is extremely delicate'.

Holyoake's exhortation was not entirely successful. Fred A. Hamilton thought that the economic importance of tobacco growing was not fully understood at government level and that there seemed to be a desire 'to feed a foreign country and go without ourselves for the sake of revenue'. R.F. (Dick) Selby-Bennetts raised an appreciative laugh when, in his defence of using more local leaf and raising the duty on American leaf, he concluded, 'America did not win the war'. The meeting was wholeheartedly against the imposition of punitive excise tax and unanimously supported the proposal to send a substantial deputation to Wellington to present their case.

Other speakers supplied a variety of perspectives in the debate. D.R. Park entered an emphatic protest on behalf of the Maori landowners who would be affected by lessening usage of their lands and falling rent returns. Mrs A.D. (Eleanor) Taylor, speaking on behalf of the Women's Division (of the Farmers' Union), urged that as many women's signatures as possible be obtained for any proposed petition, as tobacco work was one of their chief means of income. The local mayors supported any efforts to protect 'their most cherished industry', and supportive resolutions were presented from the Nelson City Council and the Nelson Provincial Progress League. Strongly worded telegrams were received from several politicians, notably Harry Atmore, the member for Nelson, and Paddy Webb, Labour Party member for West Coast. The Richmond Borough Council and the growers in Stanley Brook added their voices to the increasingly militant protest.

Many at the meeting offered money to finance the delegation. As well, an advertisement was immediately placed in the *Motueka Star-Times* seeking financial support to enable as many as possible to take the message to the government.[4] It was already clear that the delegation would be at least fifty strong and that Rupert York, the Waimea Electric Power Board chairman, Hubert Everett, and Noel Lewis, a Motueka solicitor representing the Motueka Progress League, would be amongst its members. The following week the growers' executive met to formulate its submissions. They decided that the best tactic would be to campaign against the tariffs while at the same time offering counter-proposals which 'would not lose the government one penny in revenue'. Spencer Smith was unable to join the deputation and the executive appointed Noel Lewis to speak on its behalf, confident that he would represent them well.

By now, more local and regional organisations had added their support. The Motueka Harbour Board voted to send its chairman, C.L. Harvey, to swell the ranks, conscious that any reduction in tobacco, artificial manures, and coke for kiln firing shipped over the wharf would lose them hundreds of pounds in direct revenue annually. The Riwaka branch of the Farmers' Union was financially unable to send a representative, but fully supported the growers' protest against the new duties. The delegation, described as 'the largest and most influential' ever to be sent from the Nelson district, set off for the capital on Tuesday, 24 July 1934.

The group that met with the Customs Minister, Gordon Coates, in Wellington the following day now numbered close on 100, an amazing achievement given the economic climate and the size of the population. The main thrust of the delegation's argument was that if the new duties, as recommended by the Tariff Commission, were implemented fully, the tobacco-growing industry in New Zealand would die. The counter-proposal from the growers' executive was for import duty to be fixed at 3s 6d per pound, with an added sixpence per pound for stemmed and blended tobacco. A detailed and lengthy submission was delivered by Noel Lewis, although Spencer Smith had managed to attend the meeting after all.

In an address occupying six columns of the *Motueka Star-Times*, Lewis covered virtually every point the Tariff Commission had used against the local tobacco industry, answering each with precise and apparently irrefutable logic. In attacking the Tariff Commission's case that growers were being subsidised by the government through concessions on duty, he pointed out that the formula used was extremely limited and the findings short-sighted. The duty concessions, amounting to approximately £153,000, were easily offset by the income generated and circulated within New Zealand by the tobacco-growing industry. With no apologies whatever, Lewis said the commission's report could be described as inaccurate, even ridiculous. The commission had made at least one of its statements 'in a particularly thick and murky fog' and that, given its inaccuracies, it was little wonder the report had been received with 'stunned surprise and just indignation'.

Noel Lewis used the commission's own set of criteria to show that the industry did, in fact, meet most of the conditions required for favourable consideration under protective tariffs. He contended that the commissioners had applied a limited vision to the tobacco industry. By not taking all factors into account, it had 'looked at a valuable New Zealand industry from a single and, therefore, prejudiced angle'. Speeches from leaders of other groups in the delegation followed, all stressing the role played by tobacco in the local economic structure. Though not reported by the press, the address by the Women's Division speaker, Eleanor Taylor, made a great impression on those present. In closing her remarks she exhorted the politicians not to 'sacrifice the women and children of the Nelson district for the sake of revenue'. Keith Holyoake summarised the submissions by stressing the difference between government attitudes towards tobacco growers within different countries in the empire. In countries where the industry had been established for far longer than in this country, growers were afforded a significantly greater degree of tariff protection. While tragic when related to the current situation, it might be seen as humorous that the British were at that time affording greater tariff preference to New Zealand tobacco used in Britain than our own government was prepared to grant to New Zealand leaf used in domestic manufacture.

Customs Minister Gordon Coates replied in terms which would have given those present little cause for cheer. Coates reiterated that growers must 'face the facts'. The industry required a thorough overhaul and he had hoped to see a plan for restructuring advanced by the delegation. This was a rather bitter pill for the local backbench MP, Keith Holyoake, who had consistently presented just such plans to his colleagues. Continuing his reply in language which might have come straight from tobacco company submissions to the 1930 select committee, Coates maintained that the standard of New Zealand leaf was far below that of other countries, that growers were consistently overproducing, that only certain restricted areas could produce tobacco, and that, in fact, only a 'scientific farmer' could produce the quality of leaf suitable for manufacture. Growers should realise that the government had to raise a specific amount of revenue from tobacco and at present this was £1,600,000. With sales of cut and plug tobacco gaining at the expense of cigarettes, it was obvious to Coates that the government had to shift the focus of its tariff structure.

Having admonished the deputation in this fashion, and having in no way answered the points raised so comprehensively by Lewis, Coates concluded by suggesting that representatives of the industry, together with their MP, could perhaps work together to formulate a plan for the reorganisation of the industry. With a suave political summary, he complimented Holyoake on the scale of the deputation and credited him with responsibility for the range of interests present. That he had destroyed Holyoake's claim that the tariffs came not from government policy but from the commission, must have been yet another bitter pill for the local MP. Few of those present would have left with hearts full of hope for the future of their 'cherished industry'.

Members of the deputation met with a larger group of politicians the following day. Alf Fry, representing National Company growers, and Bertie Rowling, speaking on behalf of Wills growers, both addressed this gathering and felt confident they had impressed the urgency of the situation on the political representatives. The delegation returned to Motueka with the knowledge only that they had played their strongest card and must now await the outcome. They also faced claims that the group had developed splits. It was alleged that the lengthiness of Noel Lewis's submission had prevented anybody else from speaking at the main meeting with the Minister. This was countered by the members of the delegation who had addressed MPs the following day and who were confident this had been the best approach.[5]

Growers now had to turn their attentions from political manoeuvres to the early-season tasks – preparation of seed beds and working up the ground ready for planting. By mid-August, although these preparations had begun, many growers still had no contracts with manufacturing companies. Contracts were eventually presented and signed and the season proceeded. In Riwaka, the area planted out was smaller than usual but growers hoped that good-quality leaf would maintain their incomes.[6] At the same time intensive negotiations were continuing on the Tariff Bill, with public interest and support for growers remaining high. A *Nelson Evening Mail* editorial maintained that '[t]he home market ... is New Zealand's property and ... should be preserved for New Zealanders'.[7] Gerhard Husheer publicly entered the debate, stating that the government had been misled by the Tariff Commission. He claimed that 700 farmers and possibly 3000 dependents would be deprived of their livelihood if the tobacco-growing industry were allowed to die.[8] The pressure began to have an effect.

In August 1934 Spencer Smith, Harry Manoy, Bertie Rowling, Alf Fry and James

Balck were invited to Wellington for urgent discussions with Gordon Coates. Within days, amendments to the Tariff Bill were announced – import duty on unmanufactured tobacco for cigarettes was reduced by only sixpence instead of 1s 6d, and cut and plug tobacco received a sixpenny drop instead of one shilling. Excise duty on cigarettes was adjusted accordingly, while cut tobacco excise rose by fivepence-halfpenny instead of the proposed tenpence. Duty on cigarette papers was set at a halfpenny per packet of sixty, a reduction of three-farthings on the commission's recommendation. This was not entirely satisfactory to local growers but was a significant shift from the initial proposals. Husheer remained critical, believing that the new duties discriminated against pipe smokers and favoured cigarettes. Whereas the duty on cigarettes was offset by reductions in excise, the duty on pipe tobacco was not. In addition the excise on tobacco manufactured in New Zealand had been raised while duty on imported manufactured tobacco remained the same.[9] His protests were to no avail and the amended schedule of duties stood.

Still in the midst of preparations for the coming season, growers now found the pressure was on to set up a control board for the industry. The government had obviously tired of the constant upheaval it faced, together with the conflicting demands of growers and manufacturers. Long conferences were held with the Minister of Industries and Commerce, R. Masters, who first met growers and manufacturers separately and then together. Little progress was made at these meetings. Rumours began to spread around the growing community that the conferences had ended in deadlock and that Keith Holyoake was responsible. The Growers' Association executive denied this but its statement was certainly cushioned in language which left the matter open to speculation.[10] A firmer statement was made at the association's AGM in September 1934, where Fred A. Hamilton announced that time had prevented significant progress. A further meeting had been arranged so that legislation to establish the Control Board could be brought in quickly.

Some 280 growers attended this meeting, where the matter of a unified association of growers was proposed once again. Hamilton stressed that, in the present circumstances growers should be 'banded together as firmly as possible'[11] and that such an association would have many benefits. Funds for delegations to the government would be easier to raise and the matter of crop insurance could also be better addressed through one association. Surveys of growers had shown that more than ninety per cent would join such a group. The meeting voted unanimously for a united association of growers and, despite some opposition, the proposal to name the new organisation Nelson Provincial Tobacco Growers' Association was carried.[12]

After a four-way ballot between James Balck, Noel Lewis, Cyprian Brereton and Spencer Smith, the latter was elected as chairman of the new association. Noel Lewis was elected vice-chairman and only two of those present refused to sign up for membership. Later in September a meeting of 200 growers decided that the chairman of the association should be from amongst growers themselves; Spencer Smith remained as chairman of this meeting but, as a non-grower, was not now eligible to be association chairman. Deep gratitude was expressed for the contribution made by Smith who was said to have often neglected his own legal work to attend to tobacco business.¹³ Although the meeting voted to have no elected officers of the association, a ballot later in the year saw Alf Fry selected to lead the organisation, with James Balck as vice-chairman.¹⁴

A meeting in December of the newly-formed Nelson Provincial Tobacco Growers' Association showed that members were primarily Wills growers but that they were actively seeking the membership of the air-dried growers of the National Tobacco Company. Dissension was not limited to growers for different companies, however. The chairman of the new association, Alf Fry, himself a grower for National, was challenged on why, having previously asserted that 'he was finished with the industry as regarded representation on the executive', he had now sought and gained selection as a delegate. Fry responded that 'he could please himself what he did' and that growers had requested him to stand for selection. He remained in office. The influence of Cecil Nash hovered over the new organisation – some said he was strictly neutral about the new group while others claimed that he was advising 'his' growers not to join up. Stories of 'dirty work' and 'mud-slinging' were rife amongst growers and many refrained from taking up membership of the new association until they knew more of its structure and rules.

In Tapawera and Waimea growers claimed they didn't know where they were – 'One man says one thing and the next man something different.'¹⁵ So confused were Tapawera-Waimea growers that they had sent three representatives to the meeting instead of two, each one claiming that he was the duly selected delegate. No Tapawera growers had signed up with the association but, apparently believing that there were no growers in the Waimeas, they had sent two of their own, Palmer and Curnow, as the required two representatives of Ward Five. Unfortunately for the Tapawera delegates there were indeed growers in Waimea, twenty-two of them, and they had sent S. Shuttleworth to represent them on the executive. A somewhat fractious discussion took place. The three Ward Five delegates retired from the meeting and decided amongst themselves that Curnow and Shuttleworth would be the official representatives for the meeting. It was resolved to advise the growers in Waimea and Tapawera to get together to appoint two delegates between them according to the rules of the association, rules which they said they had 'never seen and had no notion of'.¹⁶

The overriding matter for consideration was still the proposed Control Board. A further meeting of growers and manufacturers had been held on 20 September and the government now wished to know if growers wanted the Control Board to have statutory powers. Suggested powers were the control of the total quantity of tobacco to be grown, the granting of licences to grow, both for domestic consumption (under contract) and for export, as well as the fixing of grades and prices. Manufacturers had considered these proposals at a separate meeting and could not agree to all the powers proposed for the authority. They suggested that a board be set up with equal representation of growers and manufactures and that its powers be limited to issuing licences based on contracts and/or manufacturers' requirements. The companies insisted on no growing without a licence and would not agree to the board having the power to fix grades and prices. In turn, the growers' representatives would not agree to the manufacturers' proposals, which were in fact for a board restricted to issuing licences to grow, and insisted on referring the whole matter back to growers.[17]

Ward meetings were held throughout October. Growers unanimously rejected the idea of such a limited Control Board. They were determined to stand fast and insist on more powers than just licensing. Keith Holyoake spoke against this trend, advising growers to accept the manufacturers' 'terms', at least in the meantime, although he also conceded it was possible that manufacturers did not want a Control Board at all.[18] The manager of Buxtons, Fred O. Hamilton, was dubious of the whole concept, feeling that growers would be the losers – 'from past experience, where Control Boards operate, the side with the biggest influence generally gets its way'.[19] The general feeling, though, was one of puzzlement, as growers had responded to government calls for control of the industry yet no progress had been made. In an attempt to clarify the position, the association finally decided to communicate directly with individual companies, again with little apparent result.

The 1934-35 season proceeded against this backdrop of political uncertainty. Although Gerhard Husheer had indicated that up to 700 farmers were dependent on tobacco growing, the industry had shrunk markedly by 1934. Grower numbers had dropped sharply, and acreage had reduced to 1358,[20] down from 2126 in 1930-31. With the effects of the Depression lessening a little and the lack of certainty in the tobacco industry, farmers were again turning their attention to alternative land use. Both Wills and National reduced the number of their contracts for 1934-35, claiming they were overstocked. Fred O. Hamilton was sceptical: 'I am willing to make a safe bet that when [the figures] are disclosed any overstock will not be of local leaf but imported.'[21] There was still a significant number of growers, however, and their production in 1934-35 was approaching two million pounds of leaf. These growers were the basis of the domestic industry and, with many smallholders now specialising in tobacco growing, there was increasing pressure to regulate local production and remove the potential for maverick producers and processors.

Growers and manufacturers met with government representatives throughout 1935 to hammer out the mechanics of a regulatory system for the industry. Meetings were held in both Wellington and Motueka, with at least six bills being drafted before agreement was reached among the parties.[22] The resultant Tobacco Industry Bill was amended slightly at a final conference of growers and manufacturers, and was introduced to Parliament in October 1935. This was barely a month before the election that was to see a Labour government in New Zealand for the first time. The bill proposed a Tobacco Board to oversee the licensing of growers and the warranting of buyers. Annual licences would be issued to growers, specifying quantities of tobacco to be grown and area to be planted. The board would have absolute authority to grant or refuse a licence to grow. Similarly, the 'warrants to buy' to be issued to manufacturers, would be strictly controlled under the board's mandate. Introducing the bill on behalf of the Minister of Industries and Commerce (R. Masters), the Minister of Agriculture, MacMillan, stated that the bill provided stability by controlling the quantity of tobacco grown in New Zealand. This eliminated the problems of overproduction that had dogged the immediate past. He insisted there would be no suggestion of a closed corporation or price fixing – the measure was intended to 'keep the industry stable' and to prevent erratic fluctuations in local production.

The Opposition leader, Michael J. Savage, was severely critical of the way in which the board was to be constituted, the proposal that the members of the board, including growers' representatives, would all be appointed by the government coming under particular fire. 'If there was not a tragedy behind it all it would be a joke,' stated Savage in his attack. Suspicious of the power wielded by international companies, Savage claimed that the growers' representatives on the board would undoubtedly be those with contracts to grow for W.D. & W.O. Wills. The board would therefore be controlled by 'a huge monopoly'.[23] Savage was also critical of the fact that the Minister with the final say in the appointments was not, in fact, a member of the House of Representatives. Masters sat in the Legislative Council and could therefore not 'be shot at'. He (Savage) would need only two guesses as to who would influence the Minister on whom to appoint.

Keith Holyoake, in speaking to the bill, denied the opposition claim that the legislation was 'merely an election stunt'. The idea of a Control Board had been raised some years before and had been brought to a head following the deputations of growers which had approached the Minister of Customs in recent months. He stated that, while the government had done all it could to secure an earlier agreement between the interested parties, 'growers were not ready to accept it then'.[24] Holyoake expressed disappointment that the power to fix grades and prices had not been included in the board's range of duties. Such a power, he believed, would have been used only to fix the minimum grades and prices, a desirable practice to avoid bargain-basement prices for surplus tobacco. He had understood directly from the Minister that this provision would be included in the bill but had since heard that representations from at least one manufacturer and an authorised representative of a growers' organisation had resulted in the relevant words being removed from the bill.

More specifically, Holyoake maintained that the Nelson Provincial Tobacco Growers' Association remained in favour of these powers being vested in the board. Since these growers were primarily Wills growers, his implication was that the National Tobacco Company and its associated growers had lobbied against the fixing of grades and prices. This was spelt out at a meeting in the Dovedale hall, where Fred A. Hamilton told growers that Cyprian Brereton and Cecil Nash had gone to Wellington the day the bill was to be brought into the House. Hamilton claimed that Brereton had told the Minister 'he represented 300 growers or words to that effect'[25] and, on their behalf, he had stated that the grade-price proposal was unworkable. Apparently, the Minister had subsequently removed the provision from the bill before presenting it in Parliament, thereby astonishing those who had thought the matter settled.

Despite these reservations, the bill proceeded through Parliament and became the Tobacco-Growing Industry Act on 26 October 1935. The new Labour Government, elected by a Depression-weary electorate on in November, promised to bring a different approach to New Zealand primary industries. This approach was indicated by the Minister of Public Works, Bob Semple, when meeting with a deputation of tobacco growers in January 1936. Speaking on tariff protection for tobacco, Semple announced that the government would not allow the industry to be 'crucified'. He considered it 'treason to allow industries to perish and then send the dominion's money to other countries'.[26] Semple had visited the Motueka area in the early 1930s and considered that many of the local farmers would have been 'out of business without tobacco'.[27] The mood of the new government was therefore thought to be far more supportive to growers than the previous administration.

Nominations were called for the positions on the board and both manufacturers and growers put forward more names than were required. Growers were still divided along company lines and each group was anxious to be adequately represented on the board. The Nelson Provincial Tobacco Growers' Association nominated Fred A. Hamilton, James Balck, H.G. Stringer and Bertie Rowling. A meeting of minor company growers in the Dovedale hall put forward the names of Hugh Thorn and Harry Manoy to represent the South and North Island respectively. Although they had been present at the association meeting, and their representatives had been included in the election of prospective board members, the minor company growers had been outnumbered and none of their three nominees was successful. Feeling a little vulnerable in the new environment, the minor company growers meeting also decided to send a deputation to the Minister to stress the importance of this group of growers.[28] The Pioneer Union of Tobacco Growers, representing National Tobacco Company growers, also chose a panel of its members as prospective board members.

Savage's government did not change the manner of appointment to which he had objected so strongly. It was now up to Dan Sullivan, the new Minister of Industries and Commerce, to decide the balance of the board's membership. In April 1936 the association sent a deputation to Wellington to discuss grower representation, stressing to the Secretary of the Department of Industries and Commerce, Louis J. Schmitt, that grower members of the board should be bona fide growers of good standing. The shortlist for the positions was Nolan Rowling (Pioneer Union), James Balck (Association), Hugh Thorn (Minor Companies), and Bertie Rowling (Association). Fred A. Hamilton (also Association) was described as 'the Minister's appointee',[29] an odd term since all board members were, in fact, to be appointed by the Minister. The composition of the first Tobacco Board was announced on 21 April 1936, with Louis Schmitt as chairman. Manufacturers were represented by Gerhard Husheer (National Tobacco Company), R.B. Smith (W. D. and H.O. Wills), K.A. Snedden (Consolidated Tobacco Company), and A.H. Spratt (General Tobacco Company). Growers' representatives were Nolan Rowling, James Balck, Hugh Thorn and the somewhat surprising appointee George Relat, an association member of Umukuri, whose name had not previously been mentioned.

The new board's chairman, Schmitt, was a former New Zealand Tourist and Trade Commissioner in Sydney, Australia. Claiming a fifteen-year involvement with the New Zealand tobacco industry,[30] he was then Secretary of the Department of Industries and Commerce. Schmitt remained board chairman for the next seventeen years. The department also provided the board with secretarial services through the appointment of Henry Leslie Wise, who was to immerse himself thoroughly in the affairs of the industry for over thirty years. Wise provided a continuity of knowledge and experience in an ever-changing political environment. In later years this counted against him in the minds of a new generation of growers who suspected his attitudes had become entrenched.

Husheer and Smith resigned from the board after their first year, while Snedden gave up his position in June 1937 with the demise of his company. Cecil Nash and E.M. Hunt replaced the former, and Snedden's place was taken by A.F. Bell, of the Nelson Tobacco Company. The cost of administering the board was partly met through a levy on raw leaf of a halfpenny per pound, to be paid by both seller and purchaser. Licences to grow and to sell tobacco leaf cost growers 3s 6d each, while manufacturers paid one shilling per 1000 pounds of leaf for a warrant to purchase New Zealand leaf. The first board members received £2 2s per day for attending meetings and could claim travelling expenses. Motueka's grower members of the board attended their first meeting in Wellington in May.[31]

With the board in place and growers now able to discuss issues of concern directly with manufacturers, all seemed set for the uncertainty and turmoil of the last few years to be put firmly into the background. Official recognition and regulation was expected to bring confidence and stability to the industry which had now survived for over ten years. The next few years, however, were to prove almost as eventful as those just left behind.

NOTES

[1] *Nelson Evening Mail*, 12.7.34.
[2] Ibid, 13.7.34.
[3] Ibid, 11.7.34.
[4] *Motueka Star-Times*, 17.7.34, 20.7.34.
[5] Ibid, 27.7.34, 10.8.34, 14.8.34.
[6] Ibid, 17.8.34, 23.11.34.
[7] *Nelson Evening Mail*, 26.7.34.
[8] Ibid, 20.7.34.
[9] Ibid, 30.8.34.
[10] Ibid, 6.9.34. Those claiming that Holyoake had influenced the discussions gave no reason for their opinion – this may simply be an indication of political divisions within the industry.
[11] Ibid,14.9.34.
[12] Ibid, 18.9.34, p.6. Cyprian Brereton moved to retain the word Motueka in the name of the group but was defeated.
[13] Ibid, 28.9.34.
[14] Ibid, 8.10.34.
[15] *Motueka Star-Times*, 4.12.34.
[16] *Nelson Evening Mail*, 4.12.34.
[17] Ibid, 28.9.34.
[18] Ibid, 8.10.34.
[19] Letter from Hamilton to J.B. Mills, Melbourne, 11.9.34, personal papers of Rona Hurley.
[20] *New Zealand Yearbook* 1935
[21] Letter from Hamilton to J.B. Mills, op cit.
[22] *Motueka Star-Times*, 26.11.35
[23] Parliamentary Debates, October 1935, pp.614-15.

24. Parliamentary Debates, October 1935, p.618.
25. *Motueka Star-Times*, 26.11.35.
26. *Nelson Evening Mail*, 13.1.36.
27. Parliamentary Debates, 1932, Vol. 234, p.744.
28. *Nelson Evening Mail*, 25.1.36.
29. Ibid, 17.4.36.
30. Ibid, 9.9.37.
31. *Nelson Evening Mail*, 8.5.36, 24.7.36.

CHAPTER TEN

Yelling for Recognition

The 1935 legislation and the formation of the Tobacco Board brought a measure of organisation to the industry. But it did not immediately solve all the ills of those attempting to make a living from tobacco growing. Grower numbers declined throughout this period, partly through the demise of speculative growing. A large shortfall between contracted poundage and actual purchases in 1935-36 saw numbers fall by more than sixty the following season. Much of this shortfall was due to severe weather damage of crops, with a widespread hailstorm in February 1936 just as the main harvest was about to begin. The storm hit crops from Riwaka to Tadmor and swung across the Dovedale area as well. Some growers lost up to two-thirds of their crop and the overall damage was estimated at more than £50,000. On 10 February Cecil Nash reported that the hail and a following frost had caused up to a 75 per cent loss, with some growers unable to harvest a single leaf.[1]

Keith Holyoake was again swift to seek help from the government. He approached the Minister of Agriculture, W. Lee Martin, who called for an immediate report on the position. Holyoake went further in his efforts this time and displayed hail-damaged apples and tobacco leaves to impress the extent of the loss on his fellow parliamentarians.[2] Holyoake's vigorous efforts on behalf of growers had earned him the parliamentary nickname of 'Minister of Tobacco'.[3] Despite Holyoake's dramatic representations, and the statement from Lee Martin that some assistance was possible, nothing more appears to have been heard on the matter.

Two years later (1938), grower numbers had fallen still further by another hundred, to 342. The Tobacco Board's primary concern was to stabilise the industry and work towards producing as much of the domestic demand for tobacco as possible. An increase in grower numbers and acreage was central to this aim. There was, however, still doubt amongst growers about the board's ability to walk a path between growers and manufacturers. This is evident in the advice to an association meeting in April 1936, immediately after the formation of the board, that grower discussions should be only a guide for their board representatives. Grower board members were reminded that they 'would probably come against arguments which were not known now'.[4] Many growers were not convinced that all the industry's ills could be addressed through a board which included manufacturers. In August 1936 a Ward Three meeting of the association recommended that a comprehensive case on the tobacco-growing industry should be prepared and presented to the Commission on Primary Industries.[5] Growers were obviously still sceptical that the government understood their position fully.

Indeed, they were still to be heard proclaiming that the board should consist only of growers and the government. Many believed that manufacturers could 'look after themselves'[6] and that the presence of company representatives on the board retarded the progress of the industry. In a letter to the *Nelson Evening Mail*, 'A Grower' claimed that the board was 'dominated by astute businessmen'[7] whose job was to protect their own interests. The Director of Horticulture for the Department of Agriculture and long-time supporter of the growing industry, J.A. Campbell, was inclined to agree, commenting that 'when it came to a question of price, growers and manufacturers were not a good combination'.[8] Two years later, the matter was still occupying growers' attention. H.G. Stringer, strongly opposed to the composition of the board, could not see why manufacturers should be on the board at all. He maintained that growers would be better off placing their concerns directly before the Minister, who could then discuss them with manufacturers.

Keith Holyoake and the grower representatives on the board were less critical of the current structure. While it was not perfect, it allowed matters to be talked over together between all parties. George Relat considered it was 'better to contest in one room than in a triangular arrangement'.[9] Many contended, with some truth, that the board itself was a triangular arrangement. In 1939 it was suggested that the board secretary should spend most, if not all, of the year based in Motueka. This would link the administration more closely to the practical realities of the industry. Growers also resented the lack of information on board activities and discussions. They felt they were cut off from the mechanism set up for the control of their industry and that the 'secrecy surrounding the board [was] quite unnecessary'.[10] The issue was to fester for many years and resurfaced in the 1950s, causing serious divisions amongst growers.

Most might have considered that the first year of the board's time would be more than fully occupied with setting up and administering the licence and warrant scheme. In fact, this first board found time to formulate a long-range plan for the development of the domestic industry, as well as turn its attention to the provision of tobacco research. The board's plan attempted to address the problems of tariff protection and bargain-basement prices for surplus tobacco. As some of these proposals included tariff and sales tax legislation, however, the government was unable to act on them until trade negotiations with Britain had been completed. Cut-price selling was addressed, with the government agreeing to introduce regulations fixing a minimum price for New Zealand tobacco leaf. For the first year this price was fixed at one shilling per pound, rising to 1s 2d the following season. This was a minimum price only and ensured that companies could not reject inferior leaf and later purchase the same leaf at very low prices. In previous seasons, prices as low as fivepence per pound had been paid.[11]

The production of relatively low-grade tobacco in areas outside Nelson complicated this issue. In New Brighton for example, the Disabled Soldiers Civil Re-establishment League had been attempting to grow tobacco for three years under extremely adverse conditions. The crop desperately needed irrigation, and the league had only a very basic kiln. The farm manager, F.J. Bromley, thought that with assistance and better equipment the area under production could be greatly extended.[12] Charles Lowe was by no means as positive in his assessment of the prospects for New Brighton. Though he arranged for a Dawn furnace to be transferred from Pongakawa, Lowe later advised Bromley that the tobacco was 'no good' and that he should move to Pongakawa.[13] On a visit to New Brighton in January 1936, the Minister of Employment, H.T. Armstrong, was similarly pessimistic on the outlook for the league and its tobacco production.[14] Poor-quality leaf from such areas could only complicate the minimum price regulations and little government support was forthcoming for any of these projects after the demise of the Pongakawa scheme in 1938-39.

With minimum prices in place, the board turned its attention to research facilities for the growing industry. A research committee was appointed with representatives from the Departments of Scientific and Industrial Research, Agriculture, and Industries and Commerce, as well as from the Cawthron Institute in Nelson. The institute was already involved in a great deal of tobacco research, including supplying 800 boxes of specially prepared seedlings to selected growers to monitor possible factors causing mosaic. This crippling tobacco disease caused mottled colouring and blistering on leaves, which sometimes appeared burnt. There was little growers could do to combat an outbreak of mosaic and affected tobacco was virtually useless. It was believed that the disease was spread through tobacco juice on the hands and clothing of workers as well as through contaminated soil. Constant attention to hygiene was stressed and growers were advised to be vigilant throughout the season. It was even suggested that machinery should be washed down after working in areas suspected of containing mosaic infection. Not all growers agreed with this assessment of mosaic as an infection. Some were convinced the disease was a 'constitutional trouble' to which the plant was specifically prone, likening it to bitter pit in apples. Justin Foster Barham, of Riwaka, found the disease greater in nursery-propagated seedlings which were delayed in their planting out. Plants he grew from seed showed no signs of mosaic at all; nor did nursery seedlings planted out immediately.[15] All agreed, however, that finding a solution would be of enormous benefit to the industry. The research planned by the Tobacco Board would be vital.

Initially the government had planned to incorporate tobacco research into existing research facilities at Palmerston North, but persistent representations from grower groups resulted in a change of heart. The research committee soon agreed to acquire about fifteen acres of land in the Motueka area and to proceed with research as quickly as possible. In 1938 the board arranged to lease seventeen acres at Umukuri and the new station was soon under way. A programme of research was already planned, with mosaic disease as the prime target. Charles Lowe applied for the post of research officer but was unsuccessful. Eighteen applications were received and J.M. Allen was appointed. Allen was from Merredin, Western Australia, and had worked as a plant pathologist for the Australian Department of Agriculture and the Commonwealth Scientific and Industrial Research Committee. Having had six years' experience in tobacco research, he described mosaic as the earliest known tobacco disease and one which was capable of causing enormous losses.[16] The Tobacco Board contributed towards the running of the station through allocating part of its levy on raw tobacco for this purpose.[17] The research station continued to play a major role throughout the history of tobacco growing in the Motueka area and remained a significant part of the board's portfolio of interests.

Charles Lowe was not without employment, however, as he was still the government instructor, now largely concerned with export tobacco. With the new board turning its attention to encouraging export ventures, Lowe was again to play a leading role. Prior to the formation of the board, Lowe had been active in Motueka in the 1935-36 season, assisting a number of growers to produce export tobacco. The initial export shipment was a mixture of Pongakawa and Motueka tobacco, graded and packed by Buxtons at its Motueka plant in King Edward Street. The upstairs floor area of the building had been doubled and specialised equipment installed to control the moisture content. Twenty-seven graders and two floor hands were employed, with Rona Hurley in charge – the factory even boasted a radio to enliven meal times.[18]

In March 1937 Lowe arrived in London with the second shipment of tobacco to sell. In his usual enthusiastic manner he inspected all facets of the classification and buying procedure in London. He found fault with the method of deciding moisture content and convinced Arthur Phillips, of Godfrey Phillips Company, to try a different method of testing he had created. Lowe claimed this established a much lower moisture content and he obtained a price of two shillings per pound, 'equal to the top price paid for American tobacco from the [sic] Newport News'.[19] Many, like Charles Lowe, never doubted that tobacco produced in the Motueka-Nelson area could compete with overseas leaf on a quality basis. It was price, or cost of production, that worked against the establishment of a viable export market for New Zealand leaf. At this stage, however, the intention was to test the London market in order to expand the opportunities for local growers, and the board fostered these initiatives. Small shipments of New Zealand leaf were shipped to London until 1939, when wartime conditions curtailed all except vital exports. For three years, however, Charles Lowe supervised the preparation of local leaf for shipment and travelled with the tobacco on occasion.

The export drive was not without problems and price was only one of these. In 1936 Fred O. Hamilton claimed that 30,000 pounds of the 1935-36 export crop had 'been boot-legged to people who should not have got it'.[20] Later in 1936 Cecil Nash made an offer, on behalf of the National Tobacco Company, to purchase export tobacco. The company's growers had produced only about half of the contracted amount and Nash was also advertising for more growers at the time. The offer was turned down, as commitments had been made to export the leaf. Replying to Nash's offer, Fred O. Hamilton felt it would be foolish to break these agreements in view of the benefit from growing for export – 500 acres had been planted for export purposes which would not otherwise have realised any tobacco income.[21] Hamilton could not resist a pointed dig at Nash in his response. Commenting that the 1935 Act had been intended, in part, to enforce contractual production obligations of growers, he hoped that growers would not be 'set a bad example by ... companies doing that which they objected to [being done] against themselves'.[22]

The unpredictable levels of production within New Zealand were creating both unwanted surpluses in some years and a significant shortages in others. Exporting might soak up surplus leaf but it could be a problem for local companies in years of shortage. In the event, the prices received for the 1936-37 shipment of leaf were slightly down on the previous year. In mid-1937 Fred O. Hamilton expressed harsh criticism of the whole idea. The leaf exported, said Hamilton, had not been the best quality. It was mainly from growers without contracts who had submitted their entire crops, thus downgrading the overall quality. In frustration, Hamilton claimed that he had advised against exporting New Zealand tobacco and he had been right.[23]

For domestic producers the picture was somewhat mixed. Although a minimum price now prevented cut-price selling, the price growers were receiving for the run of their crop was seen as inadequate. Pressures were now mounting for higher wages for workers. This would seriously increase costs of production which were already thought to be above prices being received by growers. In response to these concerns, the government instigated an inquiry into tobacco manufacturing in 1936. There seemed to be little outcome from this but some interesting points were established. A report to the inquiry from the Department of Industries and Commerce claimed that Motueka growers were responsible for 'deliberate overproduction in excess of contracts'.[24] This excess tobacco had been sold at low prices to manufacturers producing wholly or partly New Zealand tobacco products. While these prices were below the cost of production, the department stated that some growers felt it was better than nothing.

The Horticultural Division of the Department of Agriculture, in its report, elaborated on this point. In the early days of tobacco growing the National Tobacco Company had been the only firm using New Zealand tobacco and had consequently had an advantage with lower duties. New manufacturing companies, operating under the new higher duties, were forced into cost cutting to compete. They achieved this mainly through purchasing cheap leaf through 'liquidations, speculative growers and sub-standard leaf rejected by other companies'.[25] This practice encouraged growers to overproduce, or grow without a contract.

In their submissions to the inquiry, the major companies favoured a system of minimum prices and guaranteed average prices for tobacco grown under contract. Smaller companies put forward suggestions ranging from massive government subsidies to a Leaf Control Board which would act as a clearing house for domestic tobacco and which would requisition leaf from growers, thus abolishing the current contract system.[26] Growers made only minimal representations to this inquiry, and appear to have heard little in the way of concrete proposals from it. The job of growing the tobacco had to take precedence over the more esoteric niceties of how to organise the whole industry.

Growers' main concern remained the price they would receive for their harvested and cured leaf. Constant representations were made to both the Tobacco Board and the government, but the 1937-38 season ran its course with little being heard from either source. By early April 1938 no price had been set for the current crop. This delay was said to be due to the investigation taking place into costs of production, and the board suggested buying should open at 1s 8d per pound, with adjustments made later. Growers were highly dissatisfied, especially those growing for minor companies who lacked the financial back-up available to those contracted to Wills or National.[27] Growers also felt the cost of warrants was too high – those growing for both export and local use were doubly hit by these charges.

The government remained committed to the concept of exporting tobacco. In 1937 Charles Lowe returned to Motueka to oversee the establishment of a state-owned reconditioning factory. Following the demise of the Consolidated Tobacco Company, Lowe had suggested to the Department of Agriculture that the machinery and tobacco stock of the company could be purchased for the Tobacco Board for the purposes of processing for export.[28] Buxtons of Nelson expressed interest in this operation on the condition that it be appointed to handle the export leaf. Under this arrangement the government would take over the whole venture at cost if Buxtons lost its export contract.[29] The Department of Labour agreed to purchase the machinery for £400, suggesting that the government should purchase the land, buildings and plant owned by Buxtons and lease it back to the company to carry out the planned project. Motueka community interests intervened at this point and pressed for the proposed factory to be sited in the central town area. The mayor and assorted business interests, along with Charles Lowe, were successful in persuading the government to purchase two acres on the eastern side of the town,[30] behind the business area adjacent to what is now known as Deck's Reserve. The project was to be a joint effort between the Departments of Agriculture and Labour. The extent of state involvement reflects the government's commitment to creating employment opportunities in the post-Depression era.

Growers were anxious about the new facility. They considered that they should have been consulted on the whole project, and feared that the costs of the plant would be added to their costs. It was also claimed that the present growers would bear the costs of a facility designed for a far greater number.[31] Construction began, however. The layout was designed by Charles Lowe for 'maximum effect and minimum confusion'[32] and building was supervised by P. Calder. At the end of March 1938, with the factory not completed, growers expressed concern that harvested leaf would deteriorate if buying was delayed. In May Buxtons was requested to begin receiving export leaf as the new building was still not ready. Leaf was already deteriorating. If the delay continued, it could be December before all leaf was received for processing. In the event Buxtons began receiving leaf on 1 June and the new building was officially opened on 25 July 1938.

A crowd of almost 500 attended the opening. The new facility was impressive. Lowe had designed a spacious factory, purpose-built for processing export tobacco. An external verandah provided shelter for unloading in all weathers. The leaf was then parcelled into sixty-pound bundles ready for grading by the forty-one graders at tables along the south and east walls of the shed. Working by natural light from windows behind them, the graders sorted the leaf into nine government grades established by Lowe. The graded tobacco was redried in the Proctor machine and then re-moistened, taking up to forty-five minutes depending on the grade. The processed leaf was then hydraulically pressed into tierces. These were barrel-shaped containers, four feet in diameter and tapered at the top. Each tierce held approximatley 950 pounds of leaf; up to 200 tierces could be stacked at the factory ready for shipment. The plant also possessed laboratories and tanks for the manufacture of nicotine spray, presumably for use in sheep-dip products.

West Coast MP Paddy Webb, standing in for the Minister of Labour, H.T. Armstrong, declared the building officially open. Armstrong's message, read by Webb, said early growers 'grew, picked and cured ... lacking the advantage of experience ... but sustained by a spirit of enterprise'. This event was not merely the opening of a factory, he continued, but a sign that the industry had moved from the pioneering stage to an industry based on scientific principles and an assured market. Keith Holyoake described the occasion as a 'red letter day and one fraught with significance'. He hoped that expansion would soon see the factory unable to cope with the business for which it had been built. Horatio Everett referred to his initial efforts to grow tobacco fifty years before. The industry was now established and it was up to growers to produce 'a first-class article' capable of competing with other countries. The ceremony was one of positive belief in the future prospects for New Zealand tobacco. Even Fred O. Hamilton spoke on the need to expand the area under production and to remove all restrictions which limited this expansion.

In August 1938 shipment from the new factory for the export market began. In total, 171,996 pounds were sent to London, including 26,279 pounds produced at Pongakawa. Unfortunately, some of the leaf arrived in a poor condition and was returned to New Zealand where the bulk of it was sold locally. Export of New Zealand tobacco appeared less and less likely to succeed commercially. Perhaps fortuitously, external events curtailed the trials before the bald facts of the international tobacco market impacted any further. On the outbreak of the Second World War the Tobacco Board reviewed its position on exporting tobacco, and from 1939 no further New Zealand leaf was processed for export. The Deck's Reserve complex, so dear to Charles Lowe's heart, was no longer required for the purpose.

In December 1939 Lowe lost his position with the Department of Agriculture, having been government instructor since 1927. The state plant was leased to the National Tobacco Company from 1 January 1940, and despite the long history of disagreement between them, Charles Lowe sought a position at the factory. Gerhard Husheer replied, after a month's delay, that he was considering the matter. He enclosed a cheque for £25 as an advance to help Lowe out. Lowe advised Husheer that he could not keep the cheque if there was to be no job. He would return it if he had heard nothing by the end of the month. Husheer's letter in early March brought news that there was no position available for Lowe but that he could keep the money anyway. In April, however, Husheer offered Lowe a job operating the Proctor machine at the factory starting at £6 per week 'and a little bonus at the end'.[33] Charles Lowe accepted this position and continued to work at the plant until 1942.

The closing of the state reconditioning plant was to bring an unexpected benefit for National Tobacco Company growers. A significant concern for growers in the late 1930s was the method of purchasing their leaf. Wills had a small buying shed in Motueka but many growers were required to ship their leaf to the North Island for assessment. Some felt that the lack of local buying facilities allowed manufacturers to somehow manipulate prices. Though they had little influence on the outcome, growers took some comfort from being present when their tobacco was opened for inspection on the buying floor. When A.D. (Archie) Taylor claimed that discussion of local buying at an association meeting was aimed at National Tobacco, he was met with a 'loud chorus of no's'.[34] An equally vociferous chorus of 'yes!' answered Nolan Rowling when he asked if National growers would back him wholeheartedly if he proposed local buying to the board. Despite company allegiances, on this issue growers were prepared to stand together for a more satisfactory system.

Nervousness remained regarding the role and influence of companies, however. Some claimed that 'interested parties' had taken indirect action to give the impression that growers were happy to have their leaf bought elsewhere. It was felt that growers were reluctant to express their preferences openly for fear of company reprisals and that the board should be asked to conduct a secret ballot on the matter.[35] Despite these reservations, representations were made to the Tobacco Board for uniformity of buying methods to be introduced. Local buying facilities for all companies would give all growers the opportunity to be present when their leaf was inspected and allocated a grade and price. The board successfully addressed this concern. National Tobacco established local buying facilities in 1939-40 and its growers now had their leaf bought and reconditioned locally. With all leaf now purchased locally, Wills also set up a reconditioning plant in Motueka in the early 1940s. Manufacturers became an integral part of the local industrial scene.

Another area of conflict within the industry was also addressed by the board in its early days. The existence of several groups representing growers had been a source of tension since the 1920s and was believed to have influenced the final form of the 1935 Tobacco Growing Industry Act. These groups were largely based on company allegiances. In the past, personalities had also played a part in these divisions, both in the growing community and among company personnel. Many agreed with Nolan Rowling that 'what was one company's business in tobacco did not suit another'.[36] Rowling proposed a combined executive of representatives from the three groups – Nelson Provincial Tobacco Growers' Association, Nelson Pioneer Tobacco Growers' Union, and Minor Company Growers' Committee. But the feeling prevailed that, since manufacturers were represented on the board along with growers, there seemed little reason for growers to remain divided amongst themselves along company lines.

The Tobacco Board made a concerted effort to overcome the personal and organisational divisions and in 1937 the New Zealand Tobacco Growers' Federation was formed. This body incorporated all the existing growers' organisations and provided a centralised forum for growers' concerns. It also offered an outlet for those with political inclinations. Numerous members would become prominent through their roles on the federation executive and proficient in dealing with officials and politicians of all colours. This was, however, true only for men. In its entire fifty-seven-year history there were no women members of the federation executive. Nor was any woman ever proposed as a member of the Tobacco Board. Ward meetings of the federation were also largely male preserves. In Ward One many years later the chairman called for increased attendance at meetings 'including wives!'.[37] The exclamation mark in the original minutes perhaps betrays the actual attitude to this suggestion. Only in the 1970s can any record be found of women attending ward meetings, and it was the 1980s before a woman was recorded as speaking at a federation AGM. Hundreds of male farmers, of course, also took little part in the political and organisational aspects of their industry. Most were just too busy getting one annual tobacco harvest completed in time to begin the next.

To set up the federation, the Tobacco Board conducted a postal ballot. Growers were asked to decide whether to form one organisation to represent the interests of all growers. A large meeting of growers in Motueka in June 1937 had endorsed the proposal unanimously but also supported the idea of a postal ballot. The respective growers' organisations were still jockeying for position. Talks took place between them on the question of amalgamation but Cyprian Brereton, of the Nelson Pioneer Tobacco Growers' Union, representing National Tobacco Company growers, denied emphatically that any agreement to form one body had been reached. Arrangements for the ballot proceeded, however, and in July 1937 growers received their voting papers. George Relat expressed concern that the scrutineers of the ballot should not be members of the board.[38] As a grower member of the Tobacco Board, it seems he was anxious to avoid manufacturers knowing the way individual growers voted on the issue. Like Relat, James Balck, also a member of the board, cautioned secrecy. He claimed that the National Tobacco Company was against a federation of all growers.

A great deal of feeling still existed between growers and their companies and many considered the companies capable of penalising growers who opposed their views. The delay in confirming the results of the ballot did not improve matters – no announcement had been made at the time seasonal contracts were normally finalised. Many felt the companies were holding off to influence which way growers voted. By the end of July 1937 rumours were flying and growers were becoming anxious over their contract situation. Balck suggested a board meeting as soon as possible. He was advised that the next meeting would be on 11 August, although it was possible the ballot result could be announced before this to allay the rumours.[39] In the event, the result was not made known until 23 August. A majority of licenced and previous growers were in favour of a federation of all growers, 318 voting for the proposal and eighty-six against.[40]

The federation was set up on a similar ward system to that used by the former Growers' Association, with one fewer ward.[41] Each ward represented a distinct growing area. The board proposed district associations in each ward, with elected delegates forming a central executive of the federation. Equal representation on the executive for each ward was intended to prevent any one area dominating the whole federation. Growers did not pay a subscription and were members by virtue of their licences to grow tobacco. The board offered a grant of £100 towards the administration expenses of the new body and the federation was funded through the board for almost its entire history. The chairman of the Tobacco Board, Schmitt, spoke to well-attended meetings in the ward areas, explaining the board's proposals. As the board's apparent aim was for local growers to eventually produce 90 per cent of the domestic requirement, Schmitt believed it was vital for growers to 'be big enough to set aside any petty jealousies'.[42]

There was some discontent over the size of the grant for secretarial purposes. W.E. Rowling described it as a 'scandalous offer', especially as growers contributed £2 2s per acre in levies to the board. But ward meetings generally endorsed the federation proposals. Each ward elected a committee and two delegates for the federation executive. The first federation executive consisted of Fred A. Hamilton and Keith Holyoake, Ward One; R.W.S. (Dick) Stevens and V. (Vince) Davy, Ward Two;

Maurice Thorn and C.E. Jordan, Ward Three; Gilbert Eden and W.G. Kerr, Ward Four. Fred Hamilton was elected as the first chairman of the federation at the initial meeting of the executive in October 1937. Noel Lewis was unanimously selected as secretary at the same meeting. These decisions were announced to a general meeting of growers held in Bowers' Sample Rooms on 22 October 1937.

At this meeting the question of crop insurance was again raised. This time it was suggested that a levy on growers could be used to fund such a scheme. However, the majority felt that growers were already levied to the limit, and the meeting agreed to request the Tobacco Board to investigate a possible insurance scheme. The meeting also agreed to pay a chairman's honorarium of £25 per annum. Fred Hamilton assured growers that he acted at all times out of interest in the well-being of the industry and that an honorarium would make him even more conscientious on this point. There was still dissatisfaction with the composition of the Tobacco Board. A motion of no confidence was moved at an association meeting in September 1937. Amended to the effect that the association endorsed the reconstruction of the board, the resolution was referred to the new federation to deal with.[43] This may or may not have been done, but the most pressing matters for the federation were tobacco prices, crop insurance and wages for seasonal workers. Disagreement over the nature and activities of the board would have to wait.

Keith Holyoake was quickly on the job regarding prices. He delivered a strong speech in Parliament in November, reminding the Prime Minister that he had advocated the abolition of excise duty as early as 1932, yet had made no moves since coming into power. Savage replied, somewhat lamely, that 'Rome was not built in a day', whereupon Holyoake responded that growers had been waiting a lot longer than a day for some assistance from the government.[44] As part of its plan for the development of the industry the Tobacco Board arranged for a cost-of-production survey to be carried out to establish a basis for setting prices to growers. In one respect this was the beginning of a long-running problem for growers, as it laid the foundation for a cost-plus method of determining tobacco prices. This effectively shielded growers from the true value of their product to the market.

In an industry with such a wide diversity of farm types and locations, a method of assessing costs of production applicable across the board was fraught with possibilities for dispute. For many years the varying nature of production in differing districts, and on different soils, affected the cost-of-production equation. The question of the grower's margin of profit was also a frequent source of conflict. Initially, however, the survey showed justification of a minimum average price of 1s 10 1/2d per pound (1s 7 1/2d for air-dried). As manufacturers could offer only 1s 8d, the balance of this price was provided through government subsidy. The price remained at this level until the 1943-44 season, when it was raised to 1s 11 1/2d. The subsidy was recouped through an increase in import duties and a similar increase in duties on domestically manufactured tobacco products. Essentially the consumer paid to secure the local industry. With virtually guaranteed prices, and a minimum below which tobacco could not be bought, a major concern within the industry was settled, at least for a time.

Those hard at work on their farms were often unaware of the currents of political and commercial waters. For them, life changed little during the 1930s. Each year the weather delivered an unwelcome blow to a number of farmers through flood, hail or frost. The poor season of 1935-36 was followed by losses through heavy flooding the next season. In February 1937 the Motueka River broke its banks at Woodman's Corner and at the Bluegum Corner on the West Bank. Many growers in these areas had the bulk of their crop under water and kiln furnaces were put out by the rising flood. Bertie Rowling estimated that only ten of the 250 acres in his district were showing above the flooding river. Horatio Everett had water seven feet deep in one paddock, with trout left swimming in pools amongst the tobacco as the water receded. Riwaka was cut off from Motueka and four feet of water flowed across the road at the Jubilee Bridge in Lower Moutere.[45] On the positive side, technology was bringing changes to the way in which tobacco was handled. Oil-fired kilns were introduced during 1937, although their owners found the cost of fuel oil increased by a penny per gallon in the course of the season.[46] Much of the work remained labour-intensive, although tractors were becoming more common. Mechanisation did not get its major boost until the 1940s, and many farms continued to hand-hoe and hand-tie tobacco well into the 1960s.

Wages for seasonal workers became a contentious and anxious issue during the 1936-37 season. The New Zealand Workers' Union claimed that '[t]obacco workers were the worst paid workers in New Zealand and they were out to improve their lot'. The union began its campaign by demanding wages for both men and women of 16s per day, double the male rate of the previous year. With workers often moving from farm to farm during the course of a day's work, it was felt that the hours should be accumulated to allow the payment of overtime after a total of eight hours worked. Growers were shocked. In an industry considered among the most risky of horticultural crops, there was little latitude for wage increases of any magnitude. With its vulnerability to climatic hazards, the crop could be 'cleaned up any night'.[47]

One newspaper correspondent thought workers would do well to get rid of the agitators in their midst and avoid 'killing the goose that lays the golden egg'. A worker responded in terms that disputed the golden egg theory.[48] Already half-way through the season, workers still did not know what wages they should be getting. Would it be 10s or 16s per day? If they collected their wages and moved on to another job they would be paid at the old rate of 8s per day. With wet weather often allowing only three days work per week, workers might receive only 25s for their trouble. The federation disputed the union's claim that tobacco workers were badly paid in relation to those in other primary industries. It stated that growers needed higher prices for their tobacco if they were to pay higher wages. For his part, Keith Holyoake maintained that the widespread use of women, at lower rates of pay, amounted to 'white slavery'.[49] He acknowledged, however, that tobacco prices would have to rise by 50 per cent to meet union demands.

The issue was thrashed out at a joint meeting in early 1937, at which the president of the New Zealand Workers' Union was present. The union was convinced that growers must be making something out of the industry or they would not be carrying on and increasing their acreage. The federation claimed it could make little progress while it waited for the government's response to the Tobacco Board's development plan. The union president, Eddy, suggested a joint approach to government. This was a particularly difficult time for growers to get away from the harvest, and the federation agreed, somewhat surprisingly, that Eddy should represent them in the planned deputation.[50] Wages for tobacco workers were raised in the late 1930s as part of the restoration of pre-1931 levels of pay for many workers. Growers did as they had always done – either worked together to cut the number of outside workers needed, moved towards mechanisation, or accepted the inevitable and paid up.

Despite labour difficulties and the demise of the export initiative, the tobacco-growing industry was in good heart by the outbreak of the war. The combined effects of legislation and a radically altered government philosophy had rescued growers from the shambles and turmoil of the early 1930s. By 1940 tobacco was one of the leading cash crops of the district. Grower numbers had stabilised at 381, prices were held by government subsidy, and import restrictions looked set to favour an increase in domestic production. Kiln drying was now the major method of production, accounting for almost 90 per cent of the 1937-38 crop.[51] Many National Tobacco growers had built kilns from the early 1930s onwards and both major manufacturers were now purchasing flue-cured tobacco.

Tobacco growing was almost totally centred on the Motueka-Nelson district. Other areas accounted for only a few acres by 1939, and disappeared altogether in the early 1940s. Speculative company plantations had also disappeared, with individual contracts between grower and manufacturer being the only means of producing and purchasing tobacco. The industry was now regulated, production was increasing, growers had resolved their internal differences, and prices had stabilised. Growers and manufacturers were able to gather around the board table to discuss industry issues. The future seemed assured – indeed, the 1940s would see the growing industry solidify its position. Once the war ended, a period of development and expansion began which lasted until the end of the decade.

NOTES
[1] *Nelson Evening Mail*, 10.2.36.
[2] Ibid, 18.2.36.
[3] Letter from R.W.S. Stevens to F.D. O'Flynn, federation files, 21.8.67.
[4] *Nelson Evening Mail*, 29.4.36.
[5] Ibid, 19.8.36.
[6] Ibid, 18.8.36.
[7] Ibid, 3.9.37. The style of the language in this letter suggests it may have been written by Fred A. Hamilton.
[8] Ibid, 23.4.37.
[9] Ibid, 25.7.38.
[10] Ibid, 4.8.39.
[11] Tobacco Board correspondence, 27.5.36, National Archives file Tobacco Board 1/1.
[12] *Nelson Evening Mail*, 8.1.36.
[13] Tobacco Board correspondence, op cit, 14.5.36.
[14] *Nelson Evening Mail*, 8.1.36.
[15] *Nelson Evening Mail*, 22.12.34, 23.8.37, 7.9.41.

[16] *Nelson Evening Mail*, 30.12.37.
[17] At first this was set at one-eighth of a penny per pound, which totalled £541 6s 3d in 1936-37 and £808 14s in 1937-38. Tobacco Board report, 1937, p.8.
[18] *Nelson Evening Mail*, 3.7.36, 15.9.36.
[19] Personal papers of Charles Lowe.
[20] *Nelson Evening Mail*, 29.4.36.
[21] Ibid, 11.8.36.
[22] Letter from Hamilton to Nash, 10.8.36, personal papers of Rona Hurley.
[23] *Nelson Evening Mail*, 7.5.37.
[24] Report of Department of Industries and Commerce to the Inquiry into Tobacco Manufacturers, 20.7.36, National Archives file Tobacco Board 14/10.
[25] Ibid, 14.8.36.
[26] Ibid, representations from Consolidated Tobacco Company and Nelson Tobacco Company.
[27] *Nelson Evening Mail*, 23.12.37, 30.3.38, 6.4.38, 20.5.38.
[28] Ibid, 30.6.37.
[29] Ibid, 23.8.37.
[30] Ibid, 31.1.38, 18.2.38.
[31] *Motueka Star-Times*, 7.2.38, 29.3.38.
[32] *Motueka Star-Times Times*, 28.7.38.
[33] Letter from Husheer to Lowe, 5.4.40, personal papers of Charles Lowe.
[34] *Nelson Evening Mail*, 29.4.36.
[35] Ibid, 4.8.39.
[36] Ibid, 29.4.36.
[37] Minutes of Ward One meeting, 27.4.61, federation files.
[38] Tobacco Board correspondence. Letter from Relat to the Tobacco Board, 21.6.37.
[39] Ibid, letters form James Balck to Henry Wise, 21.6.37 and 28.7.37, and Wise's reply.
[40] Tobacco Board records, 23.8.37.
[41] Ward numbers remained at four but boundaries were again adjusted and wards renumbered: Ward One: the area west of the Motueka River and north of the Alexander Bridge; Ward Two: area east of Motueka River south to the Peninsular Bridge, including Orinoco and Upper Moutere; Ward Three: area south of Wards One and Two including Dovedale, Tadmor and Motupiko; Ward Four: area east of Two and Three including Pigeon Valley, Wai-iti and all other areas in New Zealand. See Map (appendix)
[42] *Nelson Evening Mail*, 9.9.37.
[43] *Nelson Evening Mail*, 14.9.37, 25.10.37.
[44] Ibid, 3.11.37.
[45] Ibid, 26.2.37, 1.3.37.
[46] Ibid, 23.4.37.
[47] Ibid, 3.12.36.
[48] Ibid, 3.12.36, 25.2.37.
[49] Ibid, 23.4.37.
[50] Ibid, 14.1.37.
[51] Waugh, J.R., The Changing Distribution of Tobacco Growing in Waimea County, MA thesis in geography, 1962, p.52.

CHAPTER ELEVEN

Growth, or Empire Duty?

Wartime brought conflicting pressures to bear on the tobacco industry. The need to conserve empire funds and the restraints on shipping brought increasing demands for tobacco to be produced domestically. Yet wartime shortages of labour and essential materials made increased production difficult. In 1941 growers were urged to increase their acreage, with federation chairman Fred Hamilton well aware of the future implications for the industry. In his annual report in 1941 Hamilton pointed out to growers that tobacco growing was now 'established, progressive, sound, and profitable'[1] – the war effort could make the industry unassailable. He predicted confidently that the industry could provide the total domestic requirement within five years, obviously aware that the war offered the opportunity to achieve this.

This confidence was partly based on the fact that, while growers were being urged to increase production, manufacturers were also expected to use as much local leaf as possible. But acreage decreased in 1941-42 and rose only slowly during the war years. In the immediate post-war years, production and grower numbers grew rapidly. By the 1947-48 season, production had reached 4,770,827 pounds, a 115 per cent increase on 1940 figures. Grower numbers had risen to 661, an increase of 73 per cent in the same period. The changing levels of production and acreage which occurred during the 1940s were almost all directly related to wartime restrictions and post-war needs.

The major factor inhibiting wartime expansion was the lack of sufficient suitable labour. Many of those previously available for seasonal work were absorbed by the armed forces. Others, especially women, who might have taken their places were replacing full-time workers in other industries who had entered the services. The loss of growers, their sons and experienced local workers to military service worsened the problem. In 1941 one ballot alone for call-up of Nelson men included the names of eleven young growers and seventeen men described as tobacco labourers. A further ballot of youths newly turned eighteen included four tobacco workers. In 1942 men aged between thirty-five and thirty-seven, including married men, were balloted for service. Again numbers of tobacco farmers and workers called up were significant, with eleven farmers and thirteen workers on one published list.[2] This drain on the industry's resources was partly alleviated in 1942, when the Tobacco Board succeeded in securing temporary release from military service for 'bona fide growers and key men'.[3] Through this measure, some farms faced with ceasing their tobacco growing altogether were able to carry on.

In an attempt to overcome the problem of seasonal labour, the board arranged for

Territorial soldiers to be manpowered for the tobacco harvest. From the 1941-42[4] season groups of new recruits arrived by coastal trader for their first experience of the tobacco fields. Camps were set up in various locations around the district, including the Motueka Beach Camp, Richmond Showgrounds and at the Pokororo hall. The soldiers lived in tents and were transported to the farms by army vehicles often accompanied by a commanding officer. Farmers described this scheme as 'beautiful ... they arrived on time each day and left when we had finished'.[5] Wages were paid direct to the army – the soldiers received their usual daily allowance of seven shillings. In addition, growers received a subsidy of threepence per hour on the soldiers' wages. Many of the recruits found the unaccustomed work difficult as they were expected to keep up with experienced workers. While some locals recall soldiers at the weekly dances and socials, one soldier says they were too tired to take leave when it came.[6] Although now largely forgotten, the scheme did leave its mark on the district, as an army truck is reputed to have demolished the Jubilee Bridge over the Moutere Stream on the Moutere highway.

Besides the army, other labour was also manpowered, or womanpowered, to assist in the seasonal harvest of tobacco and other crops. From the 1941-42 season a National Services placement officer was stationed in the area to organise the distribution of workers. The use of the placement officer took some time to be accepted – growers were warned in 1942 that a great deal more cooperation would be expected by the department the following season. Of 400 circulars sent to growers to ascertain their labour needs the previous season, only thirteen had been returned.[7] The scheme continued, however, and numbers of young nurses and teacher trainees found themselves transported to Motueka to take on these unfamiliar tasks. Many were townies, unversed in the strictures and requirements of rural life, who found the adjustment difficult. Similarly, local farmers were faced with workers arriving dressed 'for a garden party',[8] some complete with hat and gloves. For farms around Motueka with limited accommodation, a small number of workers were housed in the hostel at the fruit dehydration factory (known locally as 'the De-Hi') in High Street. This created a further problem as the women working in the factory were being paid considerably more than those allocated to tobacco farms. Growers complained that they could not compete for workers at these rates.[9]

Women sent to more remote areas found life much more basic than those housed at the hostel. In Stanley Brook the local school buildings were used for accommodation. The group was provided with a cook but there were no washing facilities. Workers had to 'go round the farms for baths'.[10] Locals also provided the entertainment in areas where a trip to the nearest town was a matter of several hours' travel. Olga Barker played piano for many a dance and sing-song in the local hall at Stanley Brook. In other valleys similarly distant from the towns farmers banded together to harvest and grade their crops. A number of farms pooled their resources and worked in together on a rotational basis as the crops required. It was a case of 'all hands on deck', with women and children playing a vital role. Women planted, hoed, picked and loaded alongside the men. Children also took their share of these tasks, and some actually enjoyed the opportunity to stay up until midnight keeping watch on the kiln.[11]

The influx of young women, which continued to characterise the summer months in the Motueka area for many years, provided a transfusion for the district's social scene. Traditionally many women workers married into local farming families, giving the majority of tobacco families links with areas throughout New Zealand and overseas. Men coming to the area for seasonal work were less likely to marry into local farming families.

As the wartime labour shortage continued, many considered the wages paid for tobacco work to be a factor in the problem. Angry letters were exchanged in the *Nelson Evening Mail* between the secretary of the local branch of the New Zealand Workers' Union, D.E. Pope, and the federation secretary, Noel Lewis. Pope claimed that the federation's attitude in wage negotiations was driving workers away from the industry. Current wages for tobacco work were threepence per hour below the basic wage awarded by the Arbitration Court eight years before. This basic wage had since been increased several times. Lewis responded by reminding Pope that the present rates had been agreed to by the national executive of his union. Any increases for tobacco workers were dependent on the federation's efforts to get higher prices for tobacco. Accusing Pope of indulging in propaganda, Lewis suggested that the union secretary would be 'better employed in attending a tobacco executive meeting than writing harmful nonsense to the Press'. Pope replied that Lewis's reference to a wages agreement would be more relevant if he had mentioned that the negotiations had been conducted in 1943, some two years prior to the present exchange of letters.[12]

Tobacco worker B. Hopwood added a somewhat bitter perspective to the debate. From his observations 'tobacco work is not attractive'.[13] An ex-serviceman who had been manpowered to the district, Hopwood was disillusioned on finding that the work was for only five months of the year. Having served in the Middle East, he now resented being forced to work for growers whose sons (or they themselves) had been exempted from military service. He wondered sourly whether 'any of the critics [of the union's demands] are working for 18s a day with tax off and all wet days off'.[14] Now in his second season of being manpowered into the district, Hopwood cautioned workers to find out the true state of affairs before coming to the area for four to five months' work with what he clearly considered poor pay and conditions.

Women's wages, as was the case in virtually all industries, fell far behind even the allegedly low pay for men. Work for women in tobacco was divided into two grades, light and heavy work. Heavy work included all field work (mainly performed by men but not always) such as hoeing, planting, manuring and picking. The maximum women's rate was 1s 8d per hour, compared with 2s 3d for men doing the same work. A contentious issue for a time in the 1940s was whether women should in fact be employed in hoeing at all. In 1945 the union demanded 21s per day for women hoeing tobacco. The union also claimed that medical advice was against women doing this work. Aghast, the federation struggled to see the point. Some believed hoeing on sandy soil would not be difficult for women and others maintained that the work was lighter than apple picking. Although the union pressed the case by changing to a demand for a prohibition on women hoeing at all, the eventual resolution was left to the individual. Women were given the option to do hoeing, but with the curious proviso that women should not hoe tobacco in company with men.[15]

Most women were, however, employed in pricking out, pulling plants, laying plants out for planters, lateralling, topping, passing and tying leaf, grading and hanking. All these tasks were classified as light work, and were paid at the rate of 1s 4d per hour under the 1943 agreement. The hundreds of women who knelt or squatted for hours at tobacco beds, or who spent cold winter days in the grading sheds sorting leaf into the all-important grades, may well have wondered about the basis of these distinctions between heavy and light work. Whatever the merits or inequalities of the wage system, most women working for tobacco farmers were therefore earning 11s per day before tax. Little wonder that the growers felt there would be disruption in the De-Hi hostel, where the women working in the factory were being paid 14s per day.[16] Some growers were beginning to realise that wages would need to be higher if the labour problem was to be overcome. However, the opinion prevailed that growers could only accede to this if they could obtain a higher price for their leaf.

Growers of the 1940s did attempt to address the question of suitable accommodation for their workers as this was seen as a way of attracting labour. The transient workers of the Depression years in the 1930s had been relatively content to live in tents with few amenities, and the soldier labour had been given no choice. Now growers were required by law to provide at least basic living accommodation. Most settled for simple 'workers' baches' – one- or two-room dwellings which housed up to four workers and provided basic cooking facilities. The average tobacco farm of ten to twelve acres would require at least two of these baches. The shortage of materials during the war meant many could not meet the requirements and had to face the fact that workers would be placed where the accommodation was best. Communal hostels were suggested as a possible solution, with at least two required in Riwaka. Special meetings were held to explore the concept. The initial idea was to extend the De-Hi hostel and/or build others. Government assistance was seen to be justified, as the industry's production was needed and the country 'should be prepared to make a capital outlay to ensure its continuation and increase'.[17]

The hostel idea persisted for some years but eventually died through lack of cohesive action. There was little interest from other industries which might have contributed to the scheme for their own benefit. The overwhelming consensus of opinion was that growers' needs were best met by housing workers on their farms. In 1942 the government offered Public Works Department huts for the purpose but this offer came to nothing. A specially designed hut was then proposed, but at a cost of £325 each there was little likelihood of great numbers being erected and the proposal was rejected outright.[18] As materials became available following the war's end baches began to improve, although standards varied from farm to farm. Growers were advised to be more particular over cooking facilities in baches designed for women workers. Men were said to be less concerned with the standard of these facilities as 'they prefer to have their meals etc provided for them'.[19]

The other major restriction on the tobacco industry during the war was the severe shortage of building and other essential materials. A range of strategies was used to partly overcome these limitations. The shortages of both labour and materials encouraged a number of growers to increase their acreage of air-dried tobacco, since this crop required less labour input and less specialised buildings for drying. Air-dried leaf, as a percentage of the total crop, rose from 4 per cent in 1941-42 to 8 per cent in 1944-45.[20] The smaller acreages of Burley enabled a family to handle the crop themselves, thus reducing dependence on outside labour. The lack of building materials also affected this move. Galvanised iron was particularly difficult to obtain but other building materials needed in kiln construction were in short supply as well. In 1943 the stocks of corrugated iron held in the country amounted to 2500 tons, approximately half the normal stock requirements. Sheets of black steel, used for flue pipes, had been in short supply since 1941. Growers were urged to take greater care in protecting their existing pipes from corrosion; as a full set of new pipes cost in the vicinity of £20, this made economic sense as well.

Insurance companies also created difficulties for growers by insisting that half-inch mesh wire be placed over kiln pipes to prevent dry leaves settling on the pipes and catching fire.[21] As stocks of this netting were impossible to obtain, growers were often unable to comply with the conditions of their fire policies. Also difficult to obtain were such everyday items as torch batteries and alarm clocks, both essential for growers tending kilns during the night. The board applied itself to aiding in procuring supplies of these as well as bran, pollard, and fertilisers. It also financed the bulk purchase of coal for kiln furnaces for a number of years.[22] Tobacco companies also played a significant role in obtaining necessary supplies for their growers. Four of the five manufacturers then operating in New Zealand assisted in this way, as well as providing many growers with seasonal finance. Shortages of materials continued to be a restricting factor for some years after the war ended. Growers were advised in 1948 that there was no galvanised iron available at all[23] – no doubt there were many other supplies affected.

Among the measures growers used to combat the restrictions of wartime, the most effective was increased mechanisation. Tobacco culture had been highly labour-intensive right up to the war years but by 1943 the number of tractor-drawn planting machines had increased significantly. These required only three workers for planting the young plants into the open fields, rather than the large gangs of workers needed for hand planting. Similarly, fertiliser drills attached to tractors also reduced work hours. Two years later mechanical hoers were appearing in numbers. Also able to be mounted on a tractor, these enabled the grower to cultivate almost right up to harvest, eliminating much of the need for gangs of hand-hoers. Many farms, of course, remained labour-intensive. Either the cost of the new machinery was beyond their means, or the size of their operation did not justify the cost. Others preferred the old ways of working. These growers continued to hand-hoe and hand-tie well into the 1960s, just as some continued to use horses long after mechanised implements were introduced.

For some growers the greatest innovation was the use of oil to fire the kiln furnaces. The new fuel was described feelingly as 'a dream' compared to coal, coke and wood.[24] Using oil involved simply filling the tank, initially by pumping the oil from forty-four gallon drums, a far cry from the constant physical effort of stoking with wood or coal. Only those lucky enough to have automatic coal stokers had escaped this wearying task. Oil also enabled the use of automatic control, another boon for the weary farmer and family after years of manual monitoring of furnaces and temperatures.

By all these means the growing industry managed to almost double production during the early 1940s while acreage increased by 75 per cent.[25] Along with the growers' own efforts, several official initiatives were designed to stabilise and strengthen the domestic industry. Although manufacturers had used significant percentages of New Zealand leaf prior to the war, there was no mandatory requirement for them to continue this. In 1941 the government attached minimum usage conditions to manufacturers' import licences. These required companies to restrict imports of raw leaf to not more than four times the amount of domestic leaf to be used over a six-month period. This was raised to the equivalent of a mandatory domestic usage of 22.5 per cent for the following six months then to 25 per cent until the end of 1943. It was then raised to 27.5 per cent. In 1945 the percentage rose to 30 per cent, where it remained until abolished in 1981. Throughout this time manufacturers operated voluntary increases, at times using in excess of 50 per cent domestic leaf overall. Designed to both conserve overseas funds and encourage local growers, the mandatory usage appears never to have been a serious issue between growers and companies. It was only to become problematic with regard to international trade treaties such as GATT, and in times when local production could not meet the domestic demand, as in the mid-1950s.

The second official support measure involved the provision of crop insurance. The climatic hazards of frost, hail and flood had plagued tobacco growers from the first days of cultivation. Virtually every season a number of growers lost varying amounts of their crop and income through these natural disasters. In some of these years the government was approached for help, with mixed results or sometimes no results at all. In 1936 growers themselves had made enquiries on the viability of a private insurance scheme. Cyril Heaps had approached Lloyds, and a Lloyds representative, A.K. Gray, attended a meeting of the Growers' Association to discuss the possibility. However, in April 1937 the association was advised that Lloyds would be unable to provide the necessary cover.[26] The topic remained a constant point of discussion at grower meetings. A proposal was put to the Tobacco Board in 1937 but it came to nothing as the board would not agree to any insurance payouts to growers from the existing levy.

The wartime need to produce increased amounts of domestic leaf induced the government to consider the Tobacco Board's proposal for a crop insurance scheme. Without some form of security in the event of crop loss, growers were likely to succumb to the multitude of wartime difficulties and cease growing altogether. In 1944 the board moved towards setting up a Crop Insurance Reserve Fund to cover damage and loss from frost, hail and flood. Early proposals also included cover for fire damage but this was not incorporated in the 1945 scheme. A sub-committee of Fred A. Hamilton, Nolan Rowling and E.M. Hunt presented a report which the board adopted and then set about arranging government back-up for the scheme. Cabinet agreed to underwrite the fund until an adequate reserve was established from grower levies, to a maximum of five years. Ten thousand pounds was approved, subject to the board setting aside £2500 for the purpose and increasing grower levies to ensure that £1500 per year was

Part of the delegation of approximately 100 Motueka people who travelled to Wellington to petition the government for help for the industry in 1934. Pictured on the steps of Parliament buildings.

The many small farms in the Motueka area in the 1920s-30s are typified in this picture. Berries, vegetables, hops and pipfruit were supplemented by a few cattle or sheep. One or two acres of air-dried tobacco fitted well into this mix and provided a valuable cash income.

Empire T.C. Prospectus – Rothmans Archives

Empire Tobacco Company promotional brochure c1930.

Empire T.C. Prospectus – Rothmans Archives

A page from the Empire brochure c1930. Although the company claimed that its proposed plantations would be more efficient than small farms, the brochure included photographs from small Riwaka farms and pictures of the N.Z. Tobacco Co. buildings at Haumoana.

Pongakawa settlers 'pulling plants' at their tobacco beds. Note the wind shelter in the background.

The dust storm which finally destroyed the hope of viable tobacco farming in Pongakawa.

Fred A. Hamilton. Early grower and first Chairman of the N.Z. Tobacco Growers' Federation, which combined all previous growers' groups in 1938. Hamilton remained active in industry politics until the 1960s.

Henry Leslie Wise. Secretary and later chairman and secretary, of the Tobacco Board 1936–1968. A government appointee, Wise became unpopular with growers as he also served on the Price Tribunal which set the prices growers would receive for their leaf.

Louis J. Schmitt. First Chairman of the Tobacco Board set up in 1936. Schmitt remained chairman until 1953.

Tobacco was grown in some unlikely locations. Here a farmer inspects an impressive crop of Burley at Four River Flat, Murchison.

By the 1930s, kiln curing was predominant. Building materials varied – these Riwaka kilns are an example of corrugated iron construction.

Bins of harvested leaf arrive at the kiln complex to be tied and loaded into the kiln.

Harvesting and drying tobacco leaf was very labour-intensive. A typical shed gang takes a smoko break – note children who accompanied their mother from a very early age.

The cured leaf, still on sticks, is carried on stretchers to the bulk shed for storage until grading takes place when the harvest is over.

Yippee! The harvest over for another year.

Army labour was used during World War Two to fill in for farmers and workers away overseas. Here soldiers are camped at the Pokororo Hall.

Just as well it was summer. Soldiers pose at the tent flap – the fence doubles as a clothes line.

Probably the most tiresome task of all. 'Pricking out' the tiny seedlings into the seed-beds c1941.

Into the 1940s, many of the methods remained the same. Tying tobacco ready for drying still required up to nine workers.

Life wasn't all work! These 'gorgeous gussies' find another use for tobacco leaves.

Grading occupied the winter days. To the right, tobacco is 'hanked' – several leaves bound together with a supple leaf around the butts.

Sowing seed into the beds. Here the minute seed is mixed with sand to achieve an even spread c1941.

Sun-conscious workers of the 1940s dressed ready for the job ahead.

The tractor begins to replace the horse but cross cultivation is only partly mechanised c1942.

During the 1940s and 1950s many groups of Maori travelled to the Motueka district for the tobacco harvest. This group is pictured on Cyril Fry's farm at Riwaka.

W.D. and H.O. Wills new buildings under construction in Saxon St. Motueka c1947.

The original clocktower with its coloured glass panels lit at night.

Another seed-sowing method – the tiny seed is watered on from a watering can c1952.

A few weeks later the plants are hardening off in the sun prior to planting out in the field.

A Godfrey Phillips truck loaded with tobacco for processing at the Wills factory in Motueka.

A plant destroyed by Mosaic disease, the scourge of growers throughout much of the 1930s, 1940s and 1950s.

Workers accommodation came in many shapes and sizes. This Riwaka bach is typical. The Riwaka hotel can be seen in the left background.

Most seasonal workers enjoyed their stay despite the often grimy nature of the work. A sing-a-long at smoko livens up the day.

Loading kilns needed agility and sure-footedness. Kilns were up to seven racks high and the sticks were passed up to be laid along the racks above the flue pipes on the floor of the kiln.

Sticks of cured leaf are neatly stacked (bulked down) to continue the chemical changes in the tobacco leaf.

Graded leaf bundled ready for delivery to the factory.

added to the fund. The proposal was approved appreciatively by growers at ward meetings and the scheme came into being in 1945. A special provision enabled the board to pay compensation to forty-one growers who had lost part of their crop in the 1944-45 season. The average received for that year was 4.8 pence per pound, or £30 10s per grower. Levies due from growers were collated by the respective manufacturing companies and remitted to the Tobacco Board. Growers included the levy in their cost of production when formulating applications for price increases.

Payouts from the crop insurance fund were based on a sliding scale. This incorporated the costs incurred at the stage the loss was sustained, and the point of the season at which the loss occurred. In 1945 a panel of assessors was appointed to inspect damaged crops. The regulations required a panel of 'not less than sixteen persons, of whom not less than four shall be persons resident in each of the four wards'.[27] Inspections were made by an assessor, together with a manufacturer's representative and the grower involved. The assessor was required to be growing for a company different from the applicant. After assessment, a report on the damage was forwarded to the Tobacco Board. The first panel of nineteen assessors was notable for the inclusion of Rona Hurley, a woman of long experience in tobacco growing and processing. She was the only woman whose name appeared in any official capacity of the tobacco industry until Sylvia Chesney replaced Henry Wise as secretary of the Tobacco Board in 1968.[28] Payments from the crop insurance fund were never to come anywhere near the actual levels of loss but the fund proved an invaluable security blanket for the industry. The scheme at least partly hedged growers against the weather hazards which affected almost every season.

Early in the war the question of rehabilitation of returned soldiers was raised within the tobacco industry. After the loss of many skilled farmers and workers to the armed forces it was clear that, with the government's call for increased production of local leaf, it would be advantageous to assist some of these men to resettle on tobacco farms on their return from war service. The scheme was fostered by C.F. (Gerry) Skinner, the MP for Buller, and Minister of Forests and Rehabilitation. Skinner was anxious to learn the lesson of the failed rehabilitation schemes which followed the First World War. These saw numbers of former soldiers abandon farms following poor selection of land, high debt loadings and falling prices for farm products. Skinner is said to have faced opposition from Walter Nash, then Minister of Finance, who was reluctant to agree to a separate fund for rehabilitation purposes.[29]

Nevertheless the tobacco rehabilitation scheme went ahead. In 1942 the Tobacco Board recommended that the selection of suitable persons should be based on experience in tobacco production and that the land for settlement should be of a type which had proven suitable for 'the commercial production of leaf'.[30] It was further suggested that any returned servicemen who wished to take up the rehabilitation scheme should be permitted to choose their own land within these guidelines. Those with limited experience were required to serve an apprenticeship with an existing grower. Applicants then faced an interview with the Rehabilitation Board before they were approved for finance and settlement.[31] The finance terms were attractive, being 2.5 per cent over twenty-five years. The scheme proceeded slowly at first but with increasing speed as the war ended. Thirty-nine returned soldiers were placed on tobacco farms by the 1948-49 season, rising to fifty-eight by 1953.

The rehabilitation scheme created a demand for tobacco farms and gave the industry a definite post-war stimulus. In general the farms contained a reasonable ratio of suitable soils for the purpose but it seems probable that 'in the post-war boom, some people attempted to grow tobacco on unsuitable land'.[32] The federation had severe reservations about the scheme as time went on. It believed that glaring mistakes had been made as a result of lack of consultation with those involved in the growing industry. The Tobacco Board, aware of the implications of this, recommended that an experienced tobacco farmer be appointed to advise the Rehabilitation Board and Fred A. Hamilton eventually filled this position in 1949. The federation remained apprehensive, still feeling that a single representative might not be sufficient to overcome the problems.[33] Farms selected for settlement under the rehabilitation scheme were spread throughout the district from Marahau to Tapawera. Few returned men were placed in the Motueka-Riwaka area as the land was already firmly held by local families and was highly utilised and highly priced.[34] Placements were largely concentrated in the Motueka Valley, where one-third of those settled by 1949 were placed. At Marahau five of the ten tobacco farms were rehab farms. Rehab farmers were just one part of the post-war boom. The continuing need to conserve empire funds placed tobacco farmers in a position to expand production and press for a greater use of New Zealand tobacco in domestic products.

Not everyone supported the idea of products made with all, or even more, New Zealand leaf. Speaking in a parliamentary debate on the shortage of tobacco products, the MP for Tauranga, F.W. Doidge, complained of the quality of 100 per cent New Zealand tobacco. He suggested the use of more Virginian tobacco to improve the products now on sale.[35] Tobacco products were in short supply, however. This was variously attributed to labour shortages for manufacturers and to the failure of two shipments of imported leaf to arrive.[36] To overcome this factories were reported to be working extra shifts and overtime to increase their output – an expansion in local leaf production would lessen the impact of import deficiencies. The political environment supported this. There were regular proclamations of loyalty to Britain, and programmes such as the Aid for Britain Campaign were designed to enlist empire assistance in conserving funds and restoring the British economy. The Tobacco Board considered its position with regard to this concept. It agreed on measures to expand the use of locally grown tobacco within New Zealand and to encourage growers to increase both acreage and production.

Acute shortages of material, especially for kiln building, affected both the Aid for Britain Campaign and the rehabilitation scheme. It was estimated in 1947 that at least forty new kilns were required to fully meet the needs of both these projects. Even those working towards increased mechanisation struggled with the difficulty of gaining the required implements and machinery. Seeking to ease the situation, the board assisted in attempts to obtain import licences for the small tractors suitable for tobacco cultivation. It was some years, however, before most types of suitable machinery became freely available.

Despite the continuing restraint of labour and material shortages, the industry expanded rapidly in the second half of the decade. Prices rose steadily from 1944 onwards, having been static for the previous five years. Even so, tobacco growers were not entirely happy. Grower discontent had been building up for some time: dissatisfaction with the procedure by which the price was set had already become a feature of the growing industry and was to remain contentious for decades. Each year the federation was required to make application to the Price Tribunal for a price for leaf for the coming season. As early as 1942 the federation decided to base its application for an increase on changes in costs to growers since the 1939-40 season. The tribunal attempted to take into account any cost increases while maintaining price stability, but the results seldom pleased tobacco growers. In 1943 growers resolved not to sign contracts for the coming season until a higher price had been fixed. The issue was used in the 1943 election campaign. The National Party candidate for the Motueka seat, J.R. Haldane, told a meeting in the Dovedale hall that the government was holding back on an increase for growers and that some manufacturers were using the minimum price as a maximum.[37] This was hardly news to growers, who were to see no increase that year. Their resolution not to enter contracts with companies was a hollow attempt to influence matters and the board was successful in its attempts to persuade growers to sign up. The official line was that companies needed to know grower numbers in order to obtain necessary supplies for the coming season. In reality there were few options for those who wished to make a living.

Growers were not placated. Each year stronger cases were prepared for the Price Tribunal. Having responded to the government's call for increased production to the best of their ability and against some odds, growers increasingly felt aggrieved that their efforts went unappreciated and unrewarded. The increasing discontent and militancy may have become evident to the politicians. The first modest increase in 1944 (one penny per pound) was followed by regular increases throughout the decade. In 1945 it was proposed to conduct an investigation into the costs of producing tobacco as a basis for setting prices. The federation pushed hard for representation on the enquiry, believing it was impossible to establish a full picture of the growing industry without direct input from growers themselves. Riwaka grower Dick Stevens expressed reservations about the Minister's assurance that increased costs would be considered on the 'usual basis', commenting acidly that his experience of this 'usual basis' was not a happy one.[38]

The Minister of Industries and Commerce, Arnold Nordmeyer, did not agree with the federation's view. He declined to include growers in the membership of the enquiry. He did, however, give instructions that the committee would consult fully with growers 'before, during, and at the conclusion of the investigation'.[39] The committee did meet with the federation executive, in March 1946, to discuss the investigation findings. Agreeing that the structure of the investigation formed a suitable basis for calculating future prices, the federation strongly stated that the price for the previous season's crop had been inadequate. The rigours of growing tobacco justified a higher reward than 'less demanding primary pursuits'.[40] The difficulties of the yield per acre calculation were also apparent at this early stage. The committee used 1000 pounds per acre as the standard yield figure, but the federation considered this 'inapplicable because practically no grower knew the area he planted'.[41] Yield figures based on estimations of acreage planted were therefore highly unreliable as a basis for price setting.

When the Price Tribunal announced the 1945-46 price, in May 1946, there began a long-running argument on the composition of the price. The investigating committee had claimed that the great majority of growers had said they were receiving a 'good and satisfactory price'[42] for their tobacco. In response federation secretary Noel Lewis marshalled the growers' concerns in a letter to the Price Tribunal. He emphasised the particular strains of tobacco growing and the failure of the price for leaf to reflect this. The mainstay of Lewis's case was the lack of reward for such factors as 'undue physical and mental strain' resulting from the long hours of work, the possibility of climatic disaster, and the neglect of family life. This especially affected children whose parents were both required to concentrate on the harvesting and grading of tobacco. Making the point that family labour was insufficiently recognised and rewarded, Lewis remarked that '[t]he businessman, or the office employee, does not call on his wife and children for aid'.[43] Another major concern was the perception that the committee's figures appeared to reinforce the view that the two-kiln unit was essential for tobacco growing to be economic. This concerned the federation, as large numbers of one-kiln growers made up the total numbers of growers and reflected the small farm nature of the district. A move to more two-kiln units could involve the need to amalgamate farms and reduce the number of farmers. This was considered a retrograde move in terms of keeping rural families viable.

In a further point Lewis stated that, while the price of leaf had risen by 3.5 pence per pound since 1939, the wage content of leaf production had risen by 4.23 pence per pound in the same period. On wage costs alone, growers' returns had not kept pace with production costs. The dispute wound on until August, when growers, beginning their preparations for the coming season, were forced to let the matter rest. But the federation knew by now that it could never relax its efforts to secure increased profitability for growers. The yearly price applications became more and more detailed. From 1944 a panel of growers, selected from throughout the area and from differing circumstances, was used to provide full records of their cost of production to back up the federation's application. This was to become the basis of a long-running dispute between the federation and the Price Tribunal, which was not resolved until legislation changing the organisation of the industry was passed in 1974. Even the presence of Henry Wise, secretary of the Tobacco Board, on the Price Tribunal during the 1940s did not help the growers. Many were suspicious that the very opposite was the case. The letters announcing price decisions, and repudiating federation claims, were often signed by Wise.

Discontent with both the composition and functions of the board, was still evident amongst growers. There were regular calls for growers to be able to elect their representatives directly, rather than appointments being made by the Minister. Although growers conducted a ballot to illustrate their preference, the Minister was still seen to have the final say. Much confusion is evident in the annual ballots. Board members not required to stand in that particular year were nominated for the ballot and others whose names did not appear on the ballots were appointed by the Minister. Growers were apparently rightly concerned that their representatives on the board were not necessarily their current choice. At the 1945 annual general meeting of the federation, Louis Schmitt, chairman of the board, regretted that too few growers were taking part in the ballot. In reply, F.T. Holyoake claimed that growers lost interest because they knew the final decision was out of their hands.[44] Later the presence of Dick Stevens on the board, while he was also the local representative of the St James Tobacco Company, was a source of suspicion. In 1947, believing that he could not be whole-hearted in representing the interests of growers, Wards Two and Four raised the question of Stevens's eligibility to be appointed as a growers' representative on the board.[45] In 1949 Stevens resigned from the board.

Dissatisfaction was also expressed at the lack of action by the board on the matter of price. The lack of support for growers on this point was attributed to the presence of manufacturers on the board. Louis Schmitt admitted to growers, at their 1945 AGM, that the composition of the board was 'unusual' and that 'sometimes he found himself in a difficult position'. At the same meeting Henry Wise exhorted growers to look at the achievements of the industry since the advent of the board. Expressing some dissatisfaction with the way in which the minimum price was being used, Fred Hamilton agreed with Wise, to the extent that few had thought in 1935 'what splendid results would ensue'[46] from their efforts to regulate the industry. Others attending were less inclined to be wholly positive about the board. Jack Martin was supported unanimously in his suggestion that the new column in the board's report, giving actual acreage against contracted acreage, be removed. Growers again agreed that such figures were impossible to verify and were misleading and incorrect. Schmitt did his best to persuade growers that a unit cost had to be established and correct measurements were vital to this. Growers should 'chain' their areas and send a certificate verifying these to the board. No one was convinced. Martin's motion was carried without a dissenting voice and the question of acreage and yield continued to bedevil the industry for years. As well as venting their feelings on this point, growers at the 1945 meeting also vigorously disputed the need for companies to set such a great number of grades. Many felt that a premium should be paid if companies required more than four basic grades of leaf.[47] This was to be another unresolved area for the foreseeable future.

Prices did rise through the latter part of the decade. In 1949-50 growers received 2s 8d per pound for flue-cured leaf and 2s 5d for air-dried, compared with 1s 10 1/2d and 1s 7 1/2d respectively in 1939-40. Consistently claiming that wage costs were not being fully recompensed in the price-fixing process, growers felt they were unable to agree to any increases which further reduced their apparently shrinking margin of profit. The lack of an experienced workforce at times affected the quality of the finished product. Inexperienced pickers could harvest unripe leaf, thereby influencing the overall grade of the crop.[48] Although the labour supply improved gradually after the end of the war, getting sufficient seasonal labour remained an annual headache for growers. With increased acreage and more growers, the demand for workers rose higher each year. Some growers reported difficulties with the Labour Department. Stories circulated of workers being told there was no work available and of travel subsidies being refused or withheld. The federation maintained a hard line in wage negotiations. It was still resisting wage increases in 1948 when it resolved to take a strong stand and let the case go to the Arbitration Court if necessary.[49]

Company activity and involvement in the district also grew markedly during the 1940s. The decision by the National Tobacco Company to lease the state-owned redrying plant gave it a physical presence in Motueka for the first time. By 1942 Wills had also established redrying facilities in Motueka on a four-and-a-half-acre block in Saxon Street. This operation was expanded in 1948 when a substantial building was erected to house new processing machines. This building incorporated larger buying floors and a cooper's shop for the manufacture of tobacco cases. An extended boiler house was built to service the new machines. Extra storage capacity was also needed for coal supplies for the boilers, which used five tons of coal per day.[50] The completed complex was a local landmark and source of employment, and the factory's smoko whistle was audible across the surrounding urban area until its sale in 1990, after which large parts of the building were demolished and relocated by the new owners.

In 1944 National Tobacco was making moves to purchase land in the town for the erection of its own redrying factory. Cecil Nash began negotiations with the solicitor for the Thorp Estate, which owned significant tracts of land within the borough. The land the company favoured was a four-acre block located on the corner of High Street and King Edward Street – the dual access was considered very attractive from a company point of view.[51] The purchase proceeded and in 1946 construction began on a grading and forwarding store. In 1948 the building was extended, resulting in a modern complex fronted by impressive double wooden doors facing the main street, similar to the doors on the company's headquarters in Napier. An imposing clock tower was built in the corner of the grounds adjacent to the intersection of High Street and King Edward Street in 1951[52] – the corner has since been known widely as Clocktower Corner. At first the tower was decorated with red and clear glass panels, which were illuminated at night. Later a trade name replaced the coloured panels.

New companies also featured. In Nelson the Nelson Tobacco Company, founded in 1934, flourished for a time in the late 1930s and early 1940s. It produced products made entirely from locally grown leaf and owned a large tobacco-growing complex in the Motueka Valley. A fire destroyed the company's new premises, ironically located next to the fire station in Halifax Street, in 1948 and the company never recovered from the setback. The farm was subdivided and allocated to returned servicemen. Another company entered the scene during the decade. The St James Tobacco Company was set up by a Mr Wicks of London whose private company made tobacco products by royal appointment. Seeking to expand, Wicks became interested in manufacturing in New Zealand and established buying facilities on Dick Stevens' farm at Riwaka. A loyal group of about twenty growers entered contracts with the new company, produced better-than-average crops and received a penny per pound more than growers for other companies.[53] While never a major player amongst companies, St James maintained its position until company manoeuvring in the early 1960s when it was absorbed by Wills. Along with Dick Stevens, Rona Hurley played a leading role in St James, being its principal buyer and a leading grower.

Locals were very impressed with this evidence of the prosperity which tobacco was bringing to the district. And tobacco had certainly influenced the growth of the small rural centre. Two newly-built factories and the existing Buxtons complex, along with the St James buildings at Riwaka, were visible evidence of the industry. In addition, much of the commercial infrastructure was beginning to centre on the services needed by the tobacco industry and those working within it. While pip-fruit and hops were also major contributors to the local economy, the rapid development of tobacco growing and its widespread nature had made a more noticeable impact over a shorter time. In 1950 the Department of Industries and Commerce described tobacco growing as 'the largest horticultural industry in the [Nelson] province, both in acreage and total revenue'.[54]

Company expansion was not limited to Nelson, however. By 1949 the major companies had established processing factories at Otaki, Feilding and Taumarunui in addition to the large parent factories in Napier and Wellington.[55] This decentralisation was largely the result of the labour shortages which characterised the 1940s. Moving to provincial centres allowed companies to employ small-town and rural workforces, especially women. Growing operations remained concentrated in the Motueka area, however, and it was there that companies had the greatest impact on the economy. Tobacco companies were also beginning to influence the social infrastructure of the district, and the 1940s set the pattern of company support for community organisations. The war years were characterised by a huge amount of community patriotism and fundraising for the war effort. Like many other centres, Motueka formed patriotic committees and supported a number of fundraising activities. Weekly concerts were held which incorporated selected items from local personalities, community singing and dancing. Goods were donated for auction and hundreds of pounds were raised in ways which involved giving money together with building national spirit and providing entertainment and enjoyment. Tobacco companies played their part by sponsoring their share of these concerts. One such company-sponsored evening, in August 1943, raised £209 10s 2d, a record amount to that time.[56]

Gradually tobacco companies became an integral part of the community. Over the years a wide range of organisations received support and financial assistance through them. Until the new legislation banning tobacco sponsorship almost every local sports activity featured a company trophy or event. Typical is the W.D. & W.O. Wills dart board, which can still be seen in the Motueka RSA rooms. Educational and cultural activities also benefited. Company scholarships were awarded annually and support was given to a wide range of community events. This development of community involvement coincided with the growth of an almost family relationship between some growers and 'their' company. Company personnel such as C.C. (Charlie) Pethybridge of Wills, E. (Eddie) Bradley of Godfrey Phillips, and W.E. (Bill) Boyden of National (later of Rothmans) developed long-standing and close ties with growers. This represented a recognition of the mutual dependence required to produce a satisfactory crop. In 1945, after a meeting with federation representatives in Wellington, W.D. &

W.O. Wills agreed to erect a modern grading and leaf-handling facility. The company undertook to process leaf locally at competitive rates, to give growers a complete description of the grades it required, to pay an extra penny per pound in full even if the district average had been exceeded. Wills also agreed to discuss all misunderstandings between the company and growers with the intention of putting right as many grievances as possible.[57] Even if these offers were expedient for the company, they reinforced growers' ties to the company and reduced the likelihood of ongoing dissatisfaction. Such was the effectiveness of these company-grower ties that, as the years went by, growers soothed their anxieties over troubling developments within their industry with the saying 'Granny Wills will look after us'.[58] Not all growers were as fortunate with their company relationships. Those growing for the Nelson Tobacco Company experienced difficulties in obtaining payment for their leaf and some had ongoing arguments in their personal dealings with the company.[59] Eventually these growers were taken over by Wills.

Increased company activity during the 1940s changed some of the ways in which companies dealt with their growers. The once-dominant personalities of advisers such as Cecil Nash were nearing the end of their involvement with growers. Sedrick Brame, initially an adviser for Wills in the 1920s, had returned to work for the National Tobacco Company in 1942. Brame's knowledge and expertise were recognised by all who came into contact with him, many claiming he had 'forgotten more about tobacco than most people knew'. Brame had trained such advisers as Ian Hamilton who, along with Brame himself and others such as P.B. (Phil) Littlejohn and A.A. (Alan) Jury, were to maintain a continuity as a new type of field officer began to appear. Young, recently graduated with degrees in agriculture, and looking to the tobacco industry as a lifelong career, this group began to be seen in the mid-1940s. Peter Wild and A.L. (Bert) Black were amongst this initial new wave of field officers. They remained in the industry until either their company ceased operating in the area or until their retirement. The new advisers found themselves far more involved in growers' lives than they had expected. Their duties included assessing and arranging the supply of seedlings, negotiating contracts to grow and sell tobacco, and calculating the growers' fertiliser needs. They also spent up to four hours with some growers, working on the farm budget and arranging company finance for those who required it. This finance was an especial attraction to the rehab growers, who received a monthly government allowance of £22. Company funding was more generous.[60]

Growers responded to the change in adviser style with some reservation. Many were accustomed to the more authoritative manner of earlier advisers and found the technology and research-driven information they were now being offered more difficult to accept. And research was certainly playing a far greater role than ever before in tobacco growing and processing. The Tobacco Research Station at Umukuri was by now in full swing. J.M. Allan resigned in 1940 and returned to Australia. He was replaced by Robert (Bert) Thomson from the Agronomy Division at Lincoln. There was great difficulty in finding a suitably qualified scientist for the post – inquiries were made in Australia, Canada and the United States without success. Thomson had no specific experience in tobacco and was sent to Canada and the US for nine months to gain the required knowledge. For the next few years the station intensified its research into mosaic and the nutrient requirements of tobacco. In 1943 extensive comparative studies were begun of mosaic. These included comparing the results of seed-bed raising of plants against those supplied by nurseries. The use of tobacco trash in seed beds and fields was also examined, as was the effect of workers' handling of plants. The results from the latter were highly convincing. There was a drop to almost nothing in mosaic infection on farms where workers' hands were thoroughly washed between handling operations. Where tobacco trash was used in soil it was shown that mosaic infection could remain active for up to four months.

In 1945 the research station reported a marked reduction in mosaic infection at various stages of the season. The station attributed this to the increased awareness of growers on handling procedures and the need for extra care. In the Motueka area the incidence had reduced from 55.7 per cent to 9.5 per cent. From that time on the problem was largely contained, although figures varied from year to year and it was never totally eliminated. At the same time the next significant disease was beginning to appear, slowly at first but with increasing incidence as the decade progressed. Verticillium wilt was reported on one farm in 1944 and was spreading noticeably by 1947. This disease became a major worry to growers over the next decade and preoccupied the researchers into the 1980s. Black root rot also plagued growers throughout this period and was a major factor in the introduction of a new variety, Virginia Gold. This variety, while resistant to root rot, presented growers with other problems, including a high yield of what was to prove rather inferior leaf.

No amount of research could contend with the seasonal variations which continued to beset growers and upset the best-laid plans for production growth. The seasons of the 1940s were as varied as any other decade. In some years there was the full range of climatic hazards – flood, frost, drought and hail. A few seasons, such as 1939-40 and 1946-47, were mainly favourable. Others were either variable or downright calamitous: 1942 brought frosts during planting and again during harvesting; 1944-45 was abnormally cool, with less sunshine and more rain; and in 1945-46 high winds and low rainfall were a problem. Growers were hit by serious widespread flooding in the latter part of the 1947-48 season, which severely affected the amount of leaf sold. The following season's crop was affected by this late flood and also suffered unseasonal frosts and hail damage.[61]

A major project to alleviate flooding and to expand the area of usable land was proposed in 1948, when plans were under way for river protection work along the Motueka and Riwaka rivers. It was thought that the protection scheme would increase the area in production and thereby decrease the overall levies on growers. The Tobacco Board became a significant participant in the scheme, contributing £20,000, to be paid in instalments over the next twenty-five years. Growers were not unanimously in favour of this proposal. Many felt strongly that the board should consult with growers before committing funds to the project.[62] The intention to impose a rating levy was seen as reducing the government's contribution and amounting to double payment by tobacco farmers. Protests subsided, however, and contracts for the Motueka River stop-banks were let in December 1950. The board reported good progress during the summer and autumn of 1951. Some dissatisfaction was still expressed by growers, who believed that the objective of increasing the area of land under production was not being met.[63] The scheme was completed by the winter of 1955. The Tobacco Board contributed its share of the costs in yearly instalments of £1213 (later $1908 plus interest) until making its final payment in 1973-74. Surrounding land was now protected from all but the worst flooding, although still prone to a degree of surface flooding.

The crop insurance fund no doubt alleviated some of the pain from these crop losses, but by the end of the decade grower numbers had again begun to fall. There was considerable disquiet that levels of production would not continue to increase. Price-fixing methods remained a concern through this period. A section of growers began to feel that increased mechanisation, while it had eased the labour situation, had often increased costs in other ways and this was not seen to be reflected in tobacco prices.

The 1940s were, however, notable for a degree of stability, both politically and within the Growers' Federation. Fred A. Hamilton retained chairmanship of the federation throughout the decade and the Labour government remained in office until 1949. Hamilton, who had been involved with tobacco growing since the 1920s, was a major influence on the early development of the federation. A big man, with a congenial personality, he was able to relate to growers well and was noted for his ability to remember all growers by name. A natural organiser, Fred Hamilton was president or chairman of many local organisations. Not one to take a back seat, he was seldom 'just on the committee',[64] nor was he likely to be secretary or treasurer. Hamilton was a confident speaker and his sense of humour and ability to mingle with a wide range of people gave him an easy rapport with the variety of personalities within the industry. There is little doubt that his particular style of leadership influenced the federation's dealings throughout his time as chairman. Hamilton's tenure ended somewhat unfortunately. In 1950 his involvement in a local coal company led to questions on the letting of tenders for the supply of coal to growers. Having been delegated the responsibility to arrange the supply of coal for the federation, Hamilton had entered into a partnership with a local contractor in obtaining the necessary supplies. The federation received a letter of protest from Transport Nelson, a rival coal supplier. While no irregularities were suggested, the conflict of interests was not viewed favourably.[65] Hamilton's resignation as chairman was accepted, but he remained on the federation executive for several years. He continued to serve on the Tobacco Board until 1954.

The official direction was towards greater production and usage of local tobacco. As a result the overall atmosphere within the industry for much of the 1940s was positive and optimistic. Local growers were unable to meet the demand for domestic tobacco, and imports increased correspondingly. This did not cause concern as long as growers could sell all they grew. By the end of the decade, however, it became clear that growers' real returns had not improved. Indeed, in many cases, the economic position of tobacco farms had deteriorated, although this was hidden by the post-war boom until late in the 1940s.[66] By this time costs were rising. Wages had climbed steadily, and an added factor for some farmers was the renewal of Maori and church leases on their land, some of which rose by 100 per cent.[67] After a decade of comparative stability and consolidation, feelings of discontent were again rising within certain sectors of the growing community.

As well as suffering from falling economic returns, growers were still grumbling over the way in which their costs were established. Some suggested that the Farm Economics Division of Canterbury Agricultural College (Lincoln) could be used to provide a more satisfactory analysis. In addition, the absence of the chairman and secretary of the Tobacco Board from annual meetings was seen as a lack of interest in the growers concerns. Schmitt and Wise had not attended an AGM since 1947. Others criticised the board for failing to exercise its function in preventing the flood of imports which took place in the late 1940s.[68] A number of factors were thus coming together to once again stir up the currents within the industry. In 1949 the long-serving Labour government was defeated by a National Party led by Sidney Holland. The following year a new name came to prominence both on the board and within the federation. In 1950 W.C. (Fred or Freddie) Wills was elected as chairman of the federation, replacing Fred Hamilton. In the same year Fred Wills was elected as a growers' representative on the Tobacco Board. These changes would prove of major importance to the industry in the coming decade.

NOTES

1. *Nelson Evening Mail*, 11.8.41.
2. Ibid, 6.8.41, 16.9.42.
3. Tobacco Board report, 1943.
4. Waugh, J.R., The Changing Distribution of Tobacco Growing in Waimea County, MA thesis in geography, 1962, p.65.
5. Interview with Dud Eggers, 21.3.95.
6. Information from Harry Cardiff, Wellington, recruit in 2nd Field Regiment, Second World War.
7. *Nelson Evening Mail*, 18.8.42.
8. Interview with Bob Williams, 15.6.95.
9. *Nelson Evening Mail*, 12.7.45.
10. Interview with Olga and Tom Barker, Stanley Brook, 22.6.95.
11. Op cit.
12. Exchange of letters between Lewis and Pope, 18.7.45, 25.7.45.
13. Ibid.
14. Ibid.
15. New Zealand Tobacco Growers' Federation minutes, 12.10.45 and 26.11.46,
16. *Nelson Evening Mail*, 2.8.44.
17. New Zealand Tobacco Growers' Federation minutes, 7.2.47.
18. *Nelson Evening Mail*, 18.8.42.
19. New Zealand Tobacco Growers' Federation minutes, 14.2.47.
20. Tobacco Board reports, 1942, 1945.
21. *Nelson Evening Mail*, 12.8.41, 21.9.43, 4.8.44.
22. Tobacco Board reports 1942, 1943, 1944.
23. New Zealand Tobacco Growers' Federation minutes, 13.8.48.
24. Interview with former grower Russ Goodall, Dehra Doon, Riwaka, 24.4.95.
25. Tobacco Board report, 1946.
26. *Nelson Evening Mail*, 19.10.36, 9.11.36, 20.4.37.
27. Tobacco Board report, 1946, p.3.
28. Ethena M. Walker began work at the Tobacco Research Station in 1964 but is not mentioned in

official reports.
29. Interview with former grower Pat Martin, 13.6.95.
30. Tobacco Board report, 1942, p.2.
31. Interview with Pat Martin.
32. Waugh, op cit, p.72.
33. New Zealand Tobacco Growers' Federation minutes, 13.8.47, 6.5.49.
34. Waugh, op cit, p.73.
35. *Nelson Evening Mail*, 14.8.43.
36. Ibid, 14.8.43, 4.9.43.
37. Ibid, 31.8.43.
38. New Zealand Tobacco Growers' Federation minutes, 31.8.45.
39. Letter from Nordmeyer to the Tobacco Growers' Federation, federation files, 11.12.45.
40. Meeting notes, 27.3.46.
41. Letter from the Tobacco Growers' Federation to Gerry Skinner, federation files, 28.3.46.
42. New Zealand Tobacco Growers' Federation minutes, 13.5.46.
43. Letter from the Tobacco Growers' Federation to the Price Tribunal, 27.5.46, federation files. Unaware of the irony of his words, W.C. Wills, Ward Two delegate on the federation, claimed that family labour gave a false picture of profits and that the industry should be able to stand on its own feet. While it was true that from the start family labour had distorted tobacco returns, it was hardly feasible that an industry whose costs and margin of profit were decided by the Price Tribunal, and which was fundamentally protected by tariffs, could be expected to 'stand on its own feet'.
44. New Zealand Tobacco Growers' Federation minutes, 31.8.45.
45. Ibid, 14.2.47, 13.8.47.
46. *Nelson Evening Mail*, 7.9.45.
47. Ibid.
48. Tobacco Board report, 1944, p.2.
49. New Zealand Tobacco Growers' Federation minutes, 2.11.48.
50. Ibid, p.13.
51. Letters from Cecil Nash to Gerhard Husheer, 1943-44, supplied by J. Husheer.
52. Tasman District Council building permits records.
53. Information on St James supplied by John Stevens.
54. AJHR, 1950 H29 p.50.
55. AJHR, 1949 H44 p.8.
56. *Nelson Evening Mail*, 2.8.43.
57. New Zealand Tobacco Growers' Federation minutes, 14.9.45.
58. Interview with former grower Graeme Emerre, Riwaka, 24.8.96.
59. New Zealand Tobacco Growers' Federation minutes, 31.8.48, 24.6.48.
60. Interview with Bert Black, 21.6.96.
61. Department of Scientific and Industrial Research reports, *AJHR*, 1941-50.
62. New Zealand Tobacco Growers' Federation minutes, 15.6.48.
63. Tobacco Board report, 1952, p.3.
64. Notes on Fred A. Hamilton provided by his daughter, Coralie Smith, Motueka.
65. New Zealand Tobacco Growers' Federation minutes, 24.5.50.
66. Waugh, op cit, p.76.
67. New Zealand Tobacco Growers' Federation minutes, 19.4.48.
68. Ibid, 24.6.48, 13.8.48.

CHAPTER TWELVE

Battles on All Fronts

As predicted, the expansion and relative stability of the 1940s did not continue into the 1950s. Grower numbers fell by almost 40 per cent between 1947 and 1953. Numbers were still 30 per cent below the 1947 level in 1958-59. Acreage reflected this reduction, falling by over 1300 acres in the same period and recovering 500 of these by 1958-59. Areas most affected were the Motueka Valley (especially Ngatimoti), Orinoco, Tapawera, Stanley Brook and Wai-iti. In Wai-iti the drop in area was especially dramatic, reaching almost 60 per cent.[1] Production varied throughout the decade, falling one and a half million pounds between 1950-51 and 1951-52. Volume of production did not return to 1950 levels until 1958-59, when it again topped five million pounds. Growers became increasingly disillusioned. The removal of tariff protection, changes in the price-fixing process and divisions within the growing sector brought uncertainty to their lives. The marginal quality of land in the outer areas, successive climatic onslaughts and the spread of debilitating tobacco diseases added to the general malaise. The need for greater managerial intensity as tobacco farms grew in size was also being felt by the growers.[2]

This combination of factors aggravated the stress within the industry. Growers were already struggling with long working hours and labour shortages which left them physically and mentally unable to deal with other reversals.[3] As prices for other primary products rose, many decided to opt out of tobacco growing and try other avenues of farm activity. Dairy and wool prices were booming in the early 1950s and those on mixed farms, common further inland, simply switched land use to be more profitable. The move also freed such farmers from the constant labour problems and reduced the risk of losing all or part of their major income source in one climatic disaster. As growers left and were not replaced by new growers the industry shrank alarmingly. The danger arose of being unable to produce sufficient leaf to supply the minimum domestic usage requirement.

The Tobacco Board formulated a policy of encouraging expansion but in reality lacked the authority or power to implement this. In addition, the government appeared susceptible to other pressures which impacted badly on local growers. Prices offered for domestic tobacco were now seen as inadequate to reward the grower for the 'effort, anxiety and strain'[4] involved in this seasonal crop in relation to other crops. They were certainly too low to encourage new growers into the industry. Some felt this problem could be satisfactorily resolved with the cooperation of all in the industry, possibly through the Tobacco Board. Growers and manufacturers were equally represented on the board, and the third party, the government, could act as mediator. The 1950s, however, were to be a little short on cooperation.

It was also believed that certain influences were being brought to bear which prevented the grower from receiving a just reward. Throughout the 1940s growers had fought to establish an acceptable price-fixing mechanism. They felt increasingly aggrieved with the results achieved through the Price Stabilisation Tribunal, which had the final say on prices. In 1951 procedures to govern the setting of a price for domestic leaf were established. They began with a survey of the costs of production which would provide the basis of a price for tobacco leaf. This survey was carried out by the Rural Economy Division of the Department of Agriculture in collaboration with the Tobacco Growers' Federation. It was hoped that agreement would be reached on costs of production at this stage of the process. The results of the survey were then submitted to the Price Tribunal, along with the federation's annual price application. If the federation disagreed with any of the conclusions reached by the survey, representation could be made to the tribunal at that time. The Price Tribunal then considered the application, taking into account representations made by the federation, and the Department of Agriculture and 'any other relevant information which might be obtained by the tribunal from any other sources'.[5]

This provision was known as Clause Five and had been inserted on the recommendation from the Tobacco Board. The federation opposed Clause Five from the start. Predictably, such a provision fostered suspicion of undue influence by manufacturers and increased growers' distrust of the board. In addition, many felt that the industry lacked real support from the government. The federation was quick to begin building a cordial relationship with the new National government, fully conscious that their old friend Keith Holyoake was both Minister of Agriculture and Deputy Prime Minister. But it was not long before government measures destroyed any hope that this government would come to their aid. This was a major concern for an industry considered to be government-sponsored: 'if our ministry is not sympathetic toward us we have no way… of solving our problem'.[6] In an effort to combat this, the 1950s were to be peppered with delegations and deputations to ministers and departments. Few of these appeared to bring results to the liking of the federation.

Changes within the federation itself also caused concern. Tension had been rising since the election of Fred Wills as chairman in 1950, in a ballot with Wakefield grower, Sam O'Hara. In the same year Wills became a member of the Tobacco Board, replacing Martin Thorn, who retired after having been a growers' representative since 1939. Coming to the district from Canterbury in 1932, Freddie Wills had married local woman Mona Mytton and settled on land at Pangatotara, quickly becoming involved in community affairs. He was selected as the Ward Two representative on the federation executive in 1947, beginning an involvement which was to continue until 1962. Strong-minded and unwilling to compromise, Wills was totally committed to the furtherance of the tobacco industry. He had determined, by 1951, that the best way to achieve this was to increase the growers' status and power within the industry. His ideas for a growers' cooperative to handle the crop had come to nothing in the 1940s but he retained the belief that growers needed to have far greater control of their industry. Wills believed that under the prevailing set-up the manufacturers received more sympathy than growers. He therefore considered those who supported the status quo to be traitors to the growing industry.

Recalled as 'a man of small stature ticking like a time bomb',[7] Freddie Wills never failed to come out fighting when he saw a need to defend what he considered to be growers' best interests. Such pugnacity was inevitably an invitation to disagreement with those who preferred a more conservative approach, and Wills's gritty character soon brought personality clashes with other leading figures. Many were uncomfortable with his scarcely veiled criticisms of companies and his willingness to tackle issues head-on, sometimes acting first and seeking approval later. Freddie Wills, for his part, had small patience with critics. He was especially scathing of Fred Hamilton, with whom he was to have a long-running personal feud.

Wills was not the only source of conflict within the administration of the industry. At Tobacco Board level there appeared to be strong feeling against the federation, especially with regard to secretary Noel Lewis. Fred Wills claimed that at his first board meeting the main topic was 'severe criticism of a Federation official',[8] almost certainly Lewis. Some board members carried their dissatisfaction with Lewis back to the local level. Following the federation's 1950 AGM, when Fred Wills was elected chairman, there was discussion, initiated by Fred Hamilton and Dick Stevens, on the secretary's position. Several attempts were made to cap the secretarial remuneration and to have the position re-advertised.[9] The motions were all lost and Lewis remained in office.

The underlying differences remained, however. At the 1951 AGM Lewis clashed publicly with the board chairman, Louis Schmitt, who criticised growers for persisting with proposals for major changes to the control of the growing industry. Schmitt further claimed that all the items mentioned in the federation's annual report as being initiated by the federation had, in fact, been board initiatives. In a stinging attack he asserted that the board represented growers' interests and would 'look after growers if there were no Federation'.[10] Amongst others, Noel Lewis angrily refuted Schmitt's claims, insisting that growers were dissatisfied and were entitled to say so. As for the initiatives all originating with the board, Lewis claimed that the crop insurance scheme, for one, was his own idea. He added that Schmitt's attack was surprising, as the board had always been given credit where it was due. After a session of heated debate, the meeting closed with rather bizarre motions of confidence in the board and appreciation of manufacturers' activities in expanding the industry.

The open hostility within the federation erupted again in a fierce exchange between Lewis and Fred Hamilton during an executive meeting in March 1952. Hamilton alleged that board members had taken exception to the tone of recent letters from the federation, drafted by Lewis. The secretary retorted that federation letters had been twisted and misconstrued by the board secretary, Wise, for years, and later by the board chairman. Lewis maintained that a personal campaign had been waged against him, both in Motueka and in Wellington, and this had now been extended to include Fred Wills.[11] Lewis laid this campaign squarely at the feet of Fred Hamilton, who, he claimed, had taken part in board meetings largely devoted to personalities and who was said to have read portions of federation executive minutes at board meetings. These allegations were supported by other grower members of the board. K.J. (Kos) Newman stated that, in the two years since the chairmanship of the federation had changed, the atmosphere on the board had been so bad that he had seriously considered resigning. Personal animosity was the key-word – he had witnessed a '... tirade lasting for hours against the Federation and its secretary'.[12] The March executive meeting ended in a vote of no confidence in Fred Hamilton and notice of a resolution to withhold executive minutes from him in future.

Early in the decade the determination to achieve greater growers' control of the industry was to reflect these deep and bitter divisions amongst growers. Aside from personal animosities, rifts surfaced over a proposal to shift control of the industry from the Department of Industries and Commerce to the Department of Agriculture. This proposal resulted in disagreement between growers and manufacturers, and divided growers themselves. Manufacturers were opposed to moving control to the Department of Agriculture as there was a strong possibility that a producer marketing board would be set up. The department appeared to favour this type of structure for primary industries. Such an authority, which would form a body between growers and manufacturers, with control over leaf supply, quality and price, would have undermined the companies' close relationship with growers. This relationship, however cordial, placed power squarely with the manufacturer.

Many growers favoured the move. Some were disillusioned with the Department of Industries and Commerce; others could see particular advantages in being under the wing of a department specifically concerned with farming matters. Through previous dealings with Agriculture Department officers, growers felt the move would bring them a friendlier, more understanding government body whose officers were experienced in the problems of farm production. The local MP, Gerry Skinner, had previously told growers that their industry would never make progress while it remained under the control of Industries and Commerce. Going further, he suggested that growers should look at setting up their own manufacturing operation to create competition for existing manufacturers and greater control for growers.[13]

The chairman of the federation, Fred Wills, became firmly in favour of a change of departments. At ward meetings around the area in 1952 he made strongly critical statements, both of the Department of Industries and Commerce and of the Tobacco Board, of which he was a member. He maintained that the board had neglected growers for the past two years and that Industries and Commerce had nearly ruined the industry. As an example, Wills cited the wheat industry which came under the same department and was suffering as well. He was 'unaware of any person acquainted with farmers and farming'[14] within the department. Support for Freddie Wills was not unanimous, however. Fred Hamilton, for one, disagreed with proposals to change departments. All grower members of the Tobacco Board had in fact agreed in 1951 that the industry should not be moved. At that time Hamilton stated that the question to be decided was 'whether to scrap what had been built up'.[15]

Wills was not abashed by this opposition. He continued to campaign around the wards for the change. Fred Hamilton and other long-time growers were alarmed at the uncompromising nature of Wills's opinions and cracks began to appear along distinctly parochial lines. Wards Two, Three and Four endorsed the federation's stand as argued by Fred Wills. Ward One, encompassing the large number of growers in the Riwaka area, voted against the proposed change. At a 'noisy and prolonged'[16] meeting, Wills and Lewis were constantly interrupted as they attempted to explain their case. Fred Hamilton told those present that he and fellow board member J.R. (Roy) Drummond had spoken to each of the company representatives on the board. All indicated strong opposition to a change. They further commented that they 'would not issue contracts to growers'[17] if such a move were made, and would cease participating on the board. This apparent threat was sufficient to convince the majority of Ward One growers that a change of departments was not in their best interests. The meeting voted accordingly.

The debate on shifting control of the industry to the Department of Agriculture certainly rekindled growers' interest and participation in the political side of their organisation. Only thirty-one growers attended the 1950 AGM and thirty-two were at the meeting in 1951, but interest rose dramatically in the ensuing year. More than 200 growers turned up at the Masonic Hall in Motueka for the 1952 meeting. Confident of the support of three-quarters of the total growers' votes, the federation executive presented the proposal for change to the massed growers. With J.T. Watts, Minister of Industries and Commerce present, Wills as chairman again covered the reasons for a change in administration. Inviting comments, he told growers, 'This is in your hands to decide.'

As growers rose to speak, opinions were sharply divided and many factors entered the debate. Fred Hamilton denied that he had claimed manufacturers would not issue contracts, backing up his claim by reading extracts from notes taken at the meeting. Nolan Rowling objected to 'an attack on the chairman of the board',[18] protesting that neither board chairman Louis Schmitt nor secretary Henry Wise had been invited to the meeting. Schmitt had attended the 1951 meeting; Wise had not been at a federation AGM since 1947. It is not clear whether they had received invitations to the 1952 meeting.

The tenor of the arguments became more acrimonious as Noel Lewis disputed Fred Hamilton's record of the Riwaka meeting. He found it strange that Hamilton now opposed the idea of a shift, having voted for a move in the past. W.G. (Bill) Kenyon had been persuaded by Schmitt that adjustments to the present system were all that was needed. Godfrey Thomas, Jack Martin and J.L. Bruce all spoke on the advantages of a change. Martin stated that, under Industries and Commerce, 1000 acres had been taken out of production. E.H. (Ted) Bloomfield felt the government had not supported the industry as it should. Feeling there had been 'a shocking lack of interest on someone's part',[19] he claimed growers were finding the crop uneconomic and were leaving to earn regular wages elsewhere.

Feelings were running hot as the Minister addressed the meeting. His primary concern was to do the best possible job as Minister. As far as the present controversy was concerned, he could not intervene in the industry's internal disagreements. On the matter of changing departments, he commented somewhat wryly that if he were 'to consider his own convenience he would say "go somewhere else" '.[20] When put to the vote, the motion to request a transfer to the Department of Agriculture was passed by 108 votes to eighty. The Minister assured the meeting that the decision would be given full consideration and would be referred to the Minister of Agriculture, Keith Holyoake, on his return from overseas. In April of the following year Holyoake was reported as being in two minds about a transfer but little further was heard on the matter.

Whatever Holyoake's position, the issue was not resolved. Control of tobacco growing remained with the Department of Industries and Commerce and growers remained divided. They were dissatisfied with both the department and the Tobacco Board, which some felt met so infrequently – sometimes as much as six months apart – it was virtually ineffectual.[21] Others were opposed to the way their federation was representing their interests and its direction under Fred Wills. At the 1952 AGM Dick Stevens moved to delete criticisms of the Tobacco Board from the federation chairman's annual report. Freddie Wills was not without his supporters, however, and the motion was defeated.

The opposition was not to be so easily deterred. Early in 1953 a breakaway growers' group was formed – the Motueka District Tobacco Growers' Association. Friction between this group and the federation coloured the industry for the next ten years. In February 1953 a meeting of 130 growers was chaired by the mayor of Motueka, tobacco grower Walter (Wally) Eginton. He had been advised by the Minister that the chairman of the Tobacco Board, Louis Schmitt, had resigned. Eginton's assistance had been requested to restore harmony in the industry. Endorsing the view that many growers had been unhappy with the federation for some time, Eginton said that two courses had been open to those who had convened the meeting. They could have called a special meeting of the federation and attempted to replace the executive – a difficult procedure since the executive consisted of members selected by the wards. Alternatively, they could move to create a new organisation altogether. The second option had been chosen as, he claimed, results would be speedier and wrangles would be avoided. Presumably he was referring to probable wrangles at a special meeting. In reality there was to be nothing but wrangling for the foreseeable future.

On the question of concern over the decline of the industry, Eginton commented that, while it was true there were fewer growers and fewer acres, the 'real question was what were the rest producing?' Recent seasons had seen record production and he was concerned that negative comments were becoming self-fulfilling. Fred Hamilton, who was to be the interim chairman of the new group, agreed with this. He also hinted at the reasons for Schmitt's resignation. Hamilton was critical of the recent behaviour of the federation and its failure to provide adequate figures to back up price applications. Dick Stevens remarked that he had thought the difficult days of the industry were past. Now, it seemed clear that decisive action was needed to 'save the patient'.

The formation of the new association was moved by Nolan Rowling and seconded by P. Clark. A local accountant, L.W. (Steve) Leppien, was appointed as secretary-treasurer. The association decided to retain the ward system, surprisingly keeping exactly the same representation as the federation, and an executive was formed of two representatives from each of the four ward areas.[22] The meeting was assured that, given the obvious support for the new organisation, it could not fail to be recognised by the Tobacco Board. '[W]ithout a doubt moneys [sic] would come from the board to meet its expenses.'

Not surprisingly, support for the Tobacco Board and its officers was endorsed and Schmitt was to be asked to reconsider his resignation. All growers would be invited to join the association provided they resigned from the federation. In a confident mood, Hamilton commented that 'the present federation would carry on for a while, but only for a while'.

This was a call to battle for the federation. Its executive (which consistently referred to the association as 'the other group')[23] met three days later to consider the position. Predictably, Freddie Wills went onto the attack. He claimed that Eginton had converted a simple request from the Minister into a destructive attack on the federation. The Minister was now in a difficult position, said Wills, as some growers believed he had initiated the move to break away from the federation. Hamilton's remarks regarding the presentation of costs of production were described as out of date. Previously Steve Leppien had prepared these for the federation but the Rural Economy Division had now taken over this function. Wills suggested that the Minister advise the new group to submit its grievances to the federation 'after the crop was safely disposed of'.[24]

The federation executive met again on 10 March 1953. The resignations of all members were handed to the secretary to be acted upon if necessary. Roy Drummond had already resigned from the Tobacco Board, feeling he had compromised his position by attending the association meeting as an observer, although his given reason was ill-health. He withdrew his resignation shortly after. The melodrama continued. Four days later three grower members of the Tobacco Board publicly endorsed the actions of the federation. Freddie Wills, Roy Drummond, and Kos Newman all made the issue one of confidence. They would resign their positions on the board if growers voted against the federation at forthcoming ward meetings. Growers were asked not to let their personal friendships be damaged by differences within the industry: board members felt a secret ballot on the matter would avoid further bad feeling. Already the issue had caused rifts between neighbours, friends and even families.

Within a month, however, the federation's view on a ballot to clarify growers' choice of organisations had changed. The feeling in mid-April was that the association was unconstitutional and that the federation should resist any suggestions coming from such a body. As the association was basing much of its opposition on the federation's constitution and rules, the threat was now to the very basis of the federation. Set up deliberately to give all wards equal representation on the executive, the federation was now claimed to be undemocratic. The association believed that wards with larger numbers of growers should be accorded a greater representation on the executive than smaller wards. Given the overwhelming support for the association in the Riwaka area, it was clear that Ward One, the largest ward numerically, favoured a change in the rules. Pronouncing itself a non-political body, the association reinforced its opposition to a transfer of departmental control and its support for the status quo of the Tobacco Board. Other objectives, obviously also those of the federation, were the achievement of a greater margin of profit for growers and an organised approach to the provision of seasonal labour. The new group was now proceeding under a full head of steam. In mid-March it claimed a membership of 229 of the 429 registered growers.

Throughout April 1953 verbal missiles were launched and countered by each side.

At times the dispute appeared to be based more on personal disagreement and antagonism than on philosophical differences over the future direction of the industry. To some growers, the personality clashes arising from Fred Wills's chairmanship of the federation were at the heart of the matter. As in many other domains, several strong characters with conflicting views were meeting head-on. Arguments flew in all directions and at all levels. Fred Hamilton went so far as to claim that the present disharmony in the industry stemmed from the retirement in 1950 of older, more experienced members of the board and the federation,[25] a fairly blunt attack on Wills. Louis Schmitt, who had delayed his resignation until 1 July, the end of the board's year, joined the mêlée. In a circular to growers Schmitt replied to the criticisms which had been levelled at himself and the board.

Noel Lewis, who had been severely criticised for failing to offer his resignation along with the rest of the federation executive, became the first major casualty of the dispute. Lewis resigned in July 1953, having been secretary to the federation since its inception in 1938. The resignation was accepted with regret and with a fulsome expression of appreciation for his long association with the growing industry. Lewis's involvement was to resurface in the 1960s, when he would be found acting in a very different capacity. Lewis was replaced as secretary of the federation by a local accountant, J.J. (Jeff) Bradley, who was to be secretary until 1961. Schmitt's long tenure as chairman of the Tobacco Board also ended in 1953. The new chairman was R.B. Tennent, the Assistant Director General of the Department of Agriculture; the government may have seen this as a concession to growers in the matter of departmental control. Amongst growers, however, the situation was now a morass of claim and counter-claim. None of the warring organisations was making any progress towards solving the problems of those it represented. Tennent himself was not happy to accept chairmanship of the board while two growers' groups were pulling against each other. At the same time the Minister of Industries and Commerce appeared willing to recognise the association and seemed to put the onus of restoring harmony in the industry onto the federation.[26] The disorder of pre-federation days was hauntingly near.

For growers, the practicalities of their daily lives remained much the same. Materials were still often difficult to obtain, especially building supplies, and labour shortages were a recurring seasonal worry. Throughout the 1950s both the federation and the Tobacco Board employed various measures to secure sufficient workers for the tobacco harvest. Radio advertising became the major focus, although newspapers and, later in the decade, cinema were also used. Pressure was applied consistently to the Labour Department to deploy workers more effectively. Accusations surfaced regularly that department officers in other centres were discouraging workers from coming to the district.

Another consistent area of pressure was for a Labour Office, staffed by a qualified labour officer, to be opened in Motueka to deal with the seasonal labour requirements of all horticultural crops. The Department of Labour would not agree to this but compromised to the extent of authorising an agency to carry out these duties. The accounting firm of Drury and Bradley held this agency for much of the 1950s, operating from the firm's office in High Street. Growers appeared relatively satisfied with the results. Workers' fares to the Nelson area were subsidised by the Labour Department to a maximum of £4 for each worker as long as they completed at least four weeks' work. Many workers, of course, came to the area on their own initiative, and some growers arranged their own advertising, often securing workers for neighbours in this way as well.[27] The outlying valleys had the greatest difficulty in obtaining labour as the majority of workers preferred to be near the towns where there was more social life to be enjoyed. Towards the mid-1950s the labour situation improved significantly. The successful recruitment programmes were only partly responsible for this. Growers moved towards greater mechanisation or reduced their acreage to an amount they could handle without the need for large numbers of seasonal workers.

The quality of labour was another frequent talking point as larger numbers of seasonal workers arrived in the region each year. Problems of anti-social behaviour, excessive drinking and law-breaking concerned the local community as much as growers. Some believed the police should take a stronger line, others that more acceptable forms of social activity should be provided. The federation, as concerned with women workers as with men, suggested that a policewoman should be stationed in the area. The police declined the suggestion. A group of local citizens formed a welfare committee to solve some of the problems, including the lack of approved social events. Regular weekend bus trips to local points of interest were arranged and were relatively successful with those workers inclined to these pursuits. It was also believed that the seasonal nature of the work contributed to the problem: if local industries could cooperate to provide year-round employment, workers would be encouraged to settle in the area. This would eliminate the more itinerant, less dependable element of the present workforce. Whatever the prevailing view, the crop had to be harvested and there was little alternative to accepting an influx of new workers each year.

During the 1940s and 1950s increasing numbers of Maori workers came to the area. They came mainly in groups, often accompanied by an older family member who was intended to supervise the young workers. Many growers found Maori workers happier and more cooperative than others and, apart from some cultural misunderstandings, were more than satisfied with the Maori groups. Many are still recalled as exceptionally good workers. Later some growers considered that specific problems arose from young rural Maori coming unaccompanied to the area. Access to drink and unsupervised living were seen as major contributors to the social problems which were causing concern. Certainly the 1950s and 1960s were to see a growth in problems involving seasonal workers but it is not evident that Maori workers were any more prone to offending the local social mores than Pakeha or overseas visitors.[28]

The local Maori community made every effort to liaise with visiting Maori workers. Maori wardens were appointed and a centre set up where social activities could be provided in a supervised environment. In 1955 the school building from Hau School in Lower Moutere was purchased and moved to the Te Awhina Marae in Pah Street, where it was converted for use as a community centre. In addition, Maori welfare officers were appointed. At first, local kaumatua or kuia filled these roles unofficially as required. In 1959 the Department of Maori Affairs appointed Kia Riwai as Maori welfare officer for the Motueka area. Kia began by being available to Maori workers at the community centre each afternoon. Her initial brief was to '... concentrate on intellectual and cultural matters'.[29] Kia Riwai was retained in the area on a permanent basis for some years.

During the 1950s the annual influx of seasonal workers became a feature of the industry. Growers' stories of their workers are legion. The early perception of workers as needy itinerants, or conversely as almost family members, still remained for some. For others, the workers of the 1950s were a different breed altogether. Along with Maori workers came Australians and a few other overseas visitors but the majority were New Zealanders, young and looking for a break from the restrictions of mid-century, conservative urban New Zealand. Living in workers' baches and moving in groups, seasonals were often a disruptive influence on the local social scene. Hotels, dance-halls and picture theatres were favourite gathering places and many locals were unimpressed with the takeover. For the residents of the usually quiet rural township, Friday night in Motueka during the harvest season was like another world. Dressed for the occasion, workers paraded the streets, invaded shops and pubs, talked loudly and generally behaved much the same as groups of high-spirited visitors anywhere in the world. A few locals were so fascinated by the exotic goings-on that they would park in High Street just to watch. Bus companies and taxis enjoyed increased business as workers required transport, both to the farms when they arrived and to town for conviviality. The conviviality was often taken back to the farm in the form of crates of beer. Stacks of these crates could be seen on the footpath at the taxi stand as workers waited for transport to their baches.[30] Motueka's retail businesses also enjoyed the spending power of seasonal workers. Cosmetics, medicines and clothes, as well as food and drink, were all popular purchases amongst workers.

All was not harmless fun and commercial gain with regard to workers, however. The town's sole policeman developed the habit of meeting buses as they brought would-be workers into town – 'It was a case of "you can stay on the bus, you can get off" ',[31] as Sergeant Bob Smith recognised certain faces. Some growers recall police visits to their farms to seek out known offenders. They were reluctant to let their workers go and would attempt to negotiate for a full day's work before handing over the wrong-doer to the police. While not all growers harboured law-breakers, many tell even more dramatic stories of their workers. When a fire destroyed a workers' bach on Beth and Murray Heath's farm in the Motueka Valley it was feared for a time that the worker had been burnt with it. Happily the worker was found, alive and well, in a nearby bach. He had set fire to his own quarters after finding a mouse sharing them with him.[32]

Most publicity and public opinion reinforced the view that seasonal workers at times ignored the bounds of morality and law. There was, however, an underlying seam in the community which viewed women workers as fair game for local men. It was not uncommon for growers' sons and other local Romeos to be seen quietly leaving workers' baches in the early hours of the morning. The availability of young women who were themselves seeking to break the shackles of a sometimes strict upbringing also contributed to this behaviour. Resulting pregnancies often placed the women in distressing situations. Many were unwilling to let their employer know of their condition and in the outer valleys a doctor's visit was a very public event. As was the nature of the 1950s, some pregnancies resulted in marriage, perhaps to a grower's son;

other young women returned home to face their parents or kept their secret by adopting out their child. The distinction between seasonals and locals remained a feature of personal identification in the Motueka area for many years. 'Oh, no, she's not a local' was often heard in discussions of those who had stayed on after the harvest and may have spent some years in the district. To locals, the distinction was quite clear.

From a worker's perspective, tobacco work was a mixed bag. Often attracted to the area by seductive advertising which suggested that tobacco work was akin to a harvest festival, many workers found that the festival consisted of 'harvesting them'.[33] Standards of accommodation, though monitored by the Labour Department, varied widely. Some were described as 'no better than hovels'.[34] The Workers' Union reported that on one farm the men's baches were only twelve feet from the women's and were so unfit to live in that the men ate and slept in the women's accommodation.[35] Union officials were further reported as claiming that some tobacco farmers were 'so far behind the times they don't understand what a toilet roll is'.[36]

Dr E. (Ted) Bassett, travelling over the Pigeon Valley hill from Wakefield to visit patients in Dovedale and the Motueka Valley during the 1950s, found some workers living in squalor. He considered many of the growers he visited to be impoverished, deeply in debt to the tobacco companies, and working in poor conditions themselves.[37] From another perspective, it could be said that some valley families, by choice and natural inclination, lived an extremely frugal lifestyle. In some cases old run-down homes and buildings were not replaced until cash could be paid or until they simply could no longer be used. In wage negotiations the federation consistently maintained that growers could not pay higher wages unless they could recover the cost through their price application. As this was regularly cut back in the 1950s, there was little scope for agreement on wage rises. The union claimed that examination of a panel of growers' accounts 'would have revealed the existence of an industry well able to pay better wages and supply better living conditions for the workers'. The federation stood its ground, however. One attempt at negotiation was described by the union as 'about eight hours of "no, no, no" '.[38]

Despite the rhetoric, a compromise was usually reached without recourse to the Arbitration Court; wages and conditions improved slowly through the decade. The federation resisted wage rises, but growers *were* concerned with the services provided for workers. The quality of bread, the type of film being screened in the local cinema, the lack of transport services, and the need for workers to be able to purchase food supplies when they arrived in the district outside normal business hours, all generated discussion. The federation also succeeded in pressuring the local Post Office Savings Bank to increase the number of tellers available on Friday evenings, which was the only opportunity for many workers to get into town to do banking. Growers also recognised the need to keep wages on a level with the fruit industry, where the work was often seen as cleaner and more appealing. Several conferences were held with pip-fruit leaders to formulate complementary arrangements. Although conditions for tobacco workers were sometimes sub-standard and wages relatively low, most seasonal workers enjoyed their stay in the area. Many returned, often to the same farm, on a regular basis. Most hardly knew of, let alone cared about, the continuing political struggles within the industry.

The battles within the growers' ranks had, in fact, only just begun. The formation of the association in 1953 was the first formal shot in a series of skirmishes which distracted the industry and diverted the press and public for some time. Despite, or perhaps because of, the formation of the rival growers' group, the federation strengthened its position on the Tobacco Board. Following a change to the Tobacco Board regulations in 1952, the grower members of the board were now elected directly by growers. Manufacturer representatives were still formally appointed by the Minister. In 1953 Roy Drummond and Fred Hamilton were replaced by Harold Holyoake, a staunch federation man, and Sam O'Hara, the vice-chairman of the federation. The new board faced an immediate predicament. For the first time the Price Tribunal failed to grant the full price recommended by the Rural Economy Division of the Department of Agriculture, under the system agreed in 1951. This price was arrived at by a complex assessment of the costs of producing tobacco and remuneration for growers' efforts. The calculations were complicated by factors such as a fair return on equity and the unpaid labour of family members. The survey conducted by the Rural Economy Division in 1951 had attempted to set guidelines for establishing a price to growers. Some interesting and contentious points arose from this.

An average yield of 1250 pounds per acre was arrived at, based on a nine-year average. The question of price based on yield had been disturbing growers for some years and remained in dispute for the following twenty years. The tendency of the Price Tribunal to use increased yields to pressure prices downwards became a problem as advances in technology and research brought methods and varieties which raised yield per acre significantly. With prices effectively restricted by the yield factor there was little incentive to use these advances to maximum effect. The higher the yield per acre the less need the Price Tribunal saw for a price increase. It took into account the overall return to high-yield growers, rather than the return per pound, which affected all growers. A second area of conflict involved the growers' profit margin. A margin of eightpence per pound had been set by the Minister of Finance in 1938. Growers did not know what this random figure was based on nor whether they had ever received it. By 1954 the margin was said to be 1s 1d per pound which, while maintaining the same margin, was still considered too low. In addition, the 1951 survey was 'confined principally to sole growers [those growing only tobacco] producing under the best conditions'.[39] It did not, therefore, provide a true reflection of the average grower, who was likely to have a mixed income farm or less favourable conditions.

Tobacco growers felt they faced unusual expenses in producing their crop, which warranted special consideration in assessing their costs of production. For example, the provision of workers' accommodation usually included basic bedding as well as kitchen utensils and cooking equipment. Given that '[m]attresses and pillows particularly have a high mortality rate',[40] replacement costs were an almost constant factor in many farm budgets. One grower reported the need to replace eight of the ten new mattresses purchased for the previous season. The grower's own financial reward was also a point of contention. Historically, for the purposes of price fixing, growers' remuneration had been calculated on the basis of the prevailing award rate for tobacco work plus threepence per hour. In 1953-54 this was raised to fourpence halfpenny per hour. This was a step in the right direction as far as the grower was concerned but still considered to be out of proportion with the differences in skill, experience, application and outlook between growers and workers. Similarly, the growers' remuneration included a managerial allowance of thirty shillings per week. This, too, was seen as inadequate reward in view of the degree of personnel management required from growers in handling a variable, often very temporary, seasonal workforce. After spending time teaching new staff the basics of tobacco picking, tying or grading, many growers headed for work on Monday morning to find fewer than half their Friday workers still with them. Persistent labour shortages made replacement difficult. Other hazards included ensuring that boisterous or unruly workers did not cause severe damage to farm property.[41] During the war, growers could even be prosecuted for employing workers not registered for the draft.[42]

For many, the relationship with workers was filled with uncertainty. The vast majority of workers lived on the farm on which they worked and enjoyed almost family contact with their employers. This was often an added pressure, especially on tobacco women who were required to oversee the well-being of their workers. This often meant trips to the doctor or hospital at any hour, ensuring access to groceries in the days before workers had individual transport and, for some, accommodating workers in the family home. Others felt obliged to supply their workers with baking for morning and afternoon teas 'because they might not have had anything to eat since the day before'.[43] In other agricultural sectors these factors were often an accepted part of employing permanent workers or contract gangs – tobacco workers were neither. Virtually all labour needed on tobacco farms was for only a few weeks' duration and some stayed only a few days (or even hours). The provision of these services to a constantly changing variety of workers was a heavy burden on country women, most of whom worked alongside the workers, as well as managing their homes and families. In addition, many were also coping with a grower husband under stress from temperamental weather, sleepless nights, unreliable or insufficient workers and unsatisfactory returns for the crop.

Following the introduction of the new price-fixing system in 1951, a price based on the Rural Economy Division's figures and agreed to by the federation, was approved in full for the next two seasons. In 1953-54 growers were therefore shocked when the tribunal granted a price per pound of twopence less than the application. They were understandably alarmed at any further erosion of their margin. A certain section of growers believed some manufacturers had influenced the price-fixing process. While growers enjoyed mainly good relationships with their respective companies and their representatives, they were also becoming more determined that manufacturers were 'not going to intrude into their domestic affairs'.[44] This was a valid concern. For some, the company intrusion included such close control of the family's personal budget that new shoes for the children were almost a company matter. Company influence was also believed to be behind the formation of the breakaway association, although this was consistently denied. Such rumours eroded the mutual respect and tolerance which had, at least on the surface, largely prevailed in the industry since the 1930s. To many, it seemed more than coincidental that the price application was reduced in the same season as the split in the federation. The existence of two grower organisations did not strengthen growers' ability to represent their case effectively.

By August 1953 the question of membership of the opposing camps was becoming clearer. The federation produced figures which alleged that 71 per cent of growers were federation members, 19 per cent supported the association and 10 per cent were said to be neutral. At the federation's 1953 AGM, attended by 112 growers, the names of known members of the association were read out. The earlier call for personal divisions among growers to be avoided had apparently been abandoned. Prior to the AGM the federation secretary had received a bundle of 204 resignations from growers all now said to be supporters of the association. On checking, it was found that of these, ninety had subsequently rejoined the federation, twenty-nine were doubtful, and two were irregular.[45] The remaining sixty-three were those read to the meeting. Goodwill was not lost entirely, however. Following the meeting, moves towards conciliation appeared hopeful for a time, with the federation agreeing to review its rules and to meet with the association. The form of this meeting was seen as important. Any talks should be conducted in such a manner as to avoid 'losing prestige or lowering confidence'[46] among the growing community. It was therefore considered more appropriate for a deputation from the association to approach the federation rather than the two bodies meeting as equals. By November 1953 the moves had been scuppered, apparently by a circular to growers from the association, criticising the federation and reinforcing its own stand. The board chairman, Tennent, was losing patience with the situation. In March 1954 it was reported that his views on the association now coincided with those of the federation.[47] He made little further attempt to reconcile the two parties.

Other issues were becoming more pressing. Price was still the major factor concerning the growers. Throughout the early part of 1954 the federation kept up a steady stream of pressure on the Price Tribunal and the government to endorse the recommended price increase. Again, however, the application was cut back, once more by twopence per pound. It was now seriously alleged that manufacturers had provided a great deal of information to the tribunal. It was reported that all manufacturers had lodged protests over the federation's application; now even Tennent felt it was obvious that companies had tried to influence and persuade the tribunal.[48] The federation tackled the problem head-on, with direct approaches to the Price Tribunal seeking reasons behind the decision. Strong letters were sent to the Minister of Industries and Commerce, the Tobacco Board and the Department of Agriculture, making pointed reference to the use of 'Clause Five'.[49] Similar letters were sent to the members of Parliament for the Nelson and Buller electorates, and to Keith Holyoake, whose involvement in the woes of the industry in the early 1930s was not forgotten.

As in 1934, the wider community was invited to support growers in their fight. Local bodies, business groups and farmers' organisations were all approached in an attempt to illustrate the impact of tobacco income on the region. This time, however, responses were mixed and lacked the wholehearted and emotional support offered twenty years earlier. Federated Farmers expressed support and a willingness to assist if possible, but made the point that few products of direct interest to that organisation were subject to the control of the Price Tribunal. The Nelson City Council referred the federation's letter to the Nelson Provincial Progress League. The council believed that the matter was not one which fell within its province, an odd choice of words given all the circumstances. For its part, the Progress League, while keen to assist, felt unable to offer more than support in principle, an offer 'hardly sufficient to produce any change of policy on the part of the Price Tribunal'. The Waimea Electric Power Board was more positive in its response, passing a resolution of support for 'the New Zealand Tobacco Growers in their endeavour to improve their industry'.

The most disappointing response was surely that of one of the bodies from which growers might have expected the strongest support – the Motueka Borough Council. In a terse five-line letter the town clerk, Cecil Wilson, advised that the federation's letter had been received and that 'no action has been taken by this Council'.[50] The pencilled exclamation of 'No comment!!!' on the bottom of this letter sums up the feelings of the federation. It was reported that councillors had indeed had 'no comment to make' on receiving the federation's request for moral support.[51] Some may have seen significance in the current mayor, Wally Eginton, being a prominent supporter of the association. Tobacco growers would have to fight this battle virtually on their own, despite the fact that the price reduction would undoubtedly affect spending power throughout the whole district.

The strong approach adopted by the federation was endorsed by growers at ward meetings. The general disquiet was expressed in Ward Four, where the growers urged the federation to seek alternatives to the Price Tribunal for the setting of prices.[52] As protest dragged on through the winter of 1954, attitudes began to harden against the growers. Impatience was expressed at their now yearly clamour over prices and their internal disputes. The prevailing official mood had affected even the board chairman, Tennent. He told growers at their 1954 AGM that tobacco growing 'is not a depressed industry'.[53] He reminded the meeting that the price paid to growers for their tobacco had risen 1s 1d per pound in five years. While Tennent later supported the federation's case for an incentive to encourage greater production, he also declared that he 'did not feel so sorry for growers as they were for themselves'.[54] The public also aired their opinions on the division and squabbling within the industry. A correspondent to the *Nelson Evening Mail* observed that the growers were 'still scrapping away'.[55]

As the frustrating year came to a close little had changed for tobacco growers. The price protest had come to nothing and the internal fractures had not been mended. Growers had to get on with the business of sowing and planting for the coming season. The usual problems of labour shortages and possible weather hazards were compounded by disillusionment and lack of power within their own industry. Leaving the industry was not an option for all. The perception of unprofitability reduced demand for tobacco land, making it increasingly difficult to sell. As many small mixed farms were uneconomic without tobacco, there was little option but to continue. In 1955 the tribunal again rejected the full costs of production as assessed by the Rural Economy Division, reducing the application by twopence per pound once more. Again the federation went onto the attack. This time it requested a direct answer to the question of 'whether the Government wishes to have the local tobacco industry in existence or not'.[56]

By now the battle front had been widened. As well as dissatisfaction with the tribunal and the board, many were now opposing the application of the district average price scheme. It was claimed this was being manipulated by manufacturers. Some considered it was 'sheer luck to bring tobacco to the factory when a higher price was needed to achieve the average'. Many also believed that the minimum average price concept was being 'treated by manufacturers as a maximum average price',[57] depressing growers' returns. As is the nature of averages, many were, in fact, receiving below the average price for their crops, especially those growing in areas with poor or moderate soils and high exposure to climatic damage. Leaf quality and yield in these areas was often comparatively low.[58] It was pointed out that if the Rural Economy Division worked out a cost-of-production price, 'how can a grower possibly carry on when he [sic] receives *less* than that price?'.[59] The answer, of course, was that the grower in this position was producing uneconomically and should not carry on at all.

In the political and economic climate of New Zealand in the 1950s this unpalatable truth was out of place. Unemployment was virtually non-existent and other industries enjoyed subsidised or guaranteed prices and government protection. The tobacco-growing industry was among the first in New Zealand to be exposed to the chill winds of free trade, in the form of removal of tariffs on imported tobacco products in 1951. But the time had not yet come when the government would, or could, remove supportive controls altogether.

NOTES
1. Waugh, J.R., The Changing Distribution of Tobacco Growing in Waimea County, MA thesis in geography, 1962, p.79.
2. Ibid, p.82.
3. Harry Holyoake notes, c.1954, p.1, Tobacco Growers' Federation files.
4. Ibid.
5. Tobacco Board report, 1951, p.3.
6. Letter from the Tobacco Growers' Federation to the Minister of Industries and Commerce, 8.6.54.
7. Interview with Pat Martin, 16.1.96.
8. Tobacco Growers' Federation minutes, 29.8.52.
9. Ibid, 24.8.50.
10. Tobacco Growers' Federation AGM minutes, 24.8.51.
11. Tobacco Growers' Federation minutes, 14.3.52.
12. Ibid.
13. Interview with Pat Martin, 16.1.96.
14. *Nelson Evening Mail*, 25.8.52.
15. Tobacco Growers' Federation minutes, 13.4.51.
16. *Nelson Evening Mail*, 11.8.52.
17. Ibid.
18. Ibid, 1.9.52.
19. Ibid.
20. Ibid.
21. Tobacco Growers' Federation minutes, 5.12.52.
22. *Nelson Evening Mail*, 26.2.53. Representatives elected were: Ward 1: S.J. Emerre, C.E. Harvey; Ward 2: J. Manoy, H. Haycock; Ward 3: D.J. Cowin, M.H. Win; Ward 4: W.G. Kenyon, F. Gaul.
23. Tobacco Growers' Federation minutes, 10.3.53.

24. *Nelson Evening Mail*, 28.2.53.
25. Ibid, 19.3.53.
26. Tobacco Growers' Federation minutes, 4.8.53.
27. Ibid, 23.9.58.
28. Interview with a former Motueka policeman, E.J. (Jim) Gardner, Upper Moutere, 31.5.95.
29. *Nelson Evening Mail*, 5.2.59.
30. Interview with John Stevens, 6.12.95.
31. Interview with former grower Kelvin Mytton, Motueka, 9.5.95.
32. E. (Beth) Heath, Motueka, unpublished recollections.
33. *NZ Truth*, Feb. 1959.
34. Ibid.
35. *Nelson Evening Mail*, 21.2.59.
36. *NZ Truth*, Feb. 1959.
37. Interview with Dr E. Bassett, Wakefield, 31.5.95.
38. *Nelson Evening Mail*, 18.2.59.
39. Tobacco Growers' Federation working paper, 1954.
40. Tobacco Growers' Federation submission to Price Tribunal, 1953-54, federation files.
41. Op cit
42. *Nelson Evening Mail*, 10.8.41.
43. Interview with former grower K.A. (Kath) Morgan, Riwaka, 23.3.95.
44. Tobacco Growers' Federation submission to Price Tribunal, 1954-54.
45. Tobacco Growers' Federation Executive minutes, 14.8.53, 28.8.53,
46. Tobacco Growers' Federation minutes, 14.8.53.
47. Ibid, 26.3.54.
48. Ibid, 5.5.54, 10.5.54.
49. Tobacco Growers' Federation correspondence, 8.6.54, Federation files.
50. Ibid, June-July 1954.
51. *Nelson Evening Mail*, 14.7.54.
52. Ibid, 21.7.54.
53. Ibid, 23.8.54.
54. Ibid, 24.8.56.
55. Ibid, 9.8.54.
56. Tobacco Growers' Federation letter to Director of Price Control, 15.4.55, federation files.
57. *Nelson Evening Mail*, 5.12.58, 22.11.60.
58. Waugh, op cit, p.89.
59. Letter from W.H. Skipper, 30.7.54, Tobacco Growers' Federation files.

CHAPTER THIRTEEN

Changes in the Wind

To growers hard at work on their farms it hardly mattered whether the real problem was tariffs, the administration of the industry, company influence, labour shortages or the price-fixing system. The accumulation of difficulties simply made the business of growing tobacco all the more onerous. Although they were probably unaware that imports of cigarettes rose from 179,000 pounds in 1951 to 1,719,000 pounds in 1955,[1] growers felt that the industry was being squeezed almost beyond endurance. From a purely personal point of view, it seemed that a price increase or some form of incentive payment would solve many of the difficulties.

The federation devoted almost its entire energies to the question of strengthening the position of growers. It constantly sought ways to reorganise the industry with this objective. Since the 1930s the idea of a growers' cooperative or marketing body had been raised periodically and was to surface again during the troubled mid-1950s. Fred Wills's concept of a growers' cooperative to control and handle all locally produced leaf was revisited, and the idea of a growers' board was strongly supported at the federation's 1955 AGM. Neither proposal came to fruition. Growers remained divided on whether to take the risk of seriously alienating manufacturers. Without a guarantee of support from the government, growers lacked the confidence to proceed on such major changes to the structure of their industry. It may have been true that most growers '[did] not give two hoots what administration [the] industry is under or what system is in vogue',[2] but the existence of two growers' organisations was symptomatic of the pain of the industry. The crucial question dividing growers was whether to stay with the status quo or risk completely reorganising their industry. The internal dissension sapped the strength of the producing sector and may have been used to its disadvantage. Certainly Fred Wills believed that the divisions among growers were acting to the advantage of manufacturers.[3] A united front might well have achieved a greater degree of grower influence on the government, either for radical change or for a more favourable use of the status quo.

The Tobacco Board itself was virtually impotent in the face of government obduracy. Almost forlornly referring to representations to the Minister, the board's 1955 report identified several problem areas, all of which revolved around the relative unattractiveness of tobacco growing. The price system, the manner of buying and grading leaf, the need for an incentive to grow, levels of taxation and excise, and the provision of sufficient labour were the primary factors the board wished to review. Even the labour question, which would seem to be the least controversial issue to resolve, was to be vexatious year after year. It was mainly eased by the board's own efforts in financially supporting a seasonal placement office in Motueka and by the federation, supported by the board, undertaking recruitment advertising in other areas.

With its peculiar composition of growers and manufacturers, the board seemed unable to make any significant impact on the matter of price fixing. In 1955 Sam O'Hara told growers that the Tobacco Board had never been any use to them and that only at the two most recent meetings had growers' concerns been addressed seriously.[4] The board was, in fact, split on issues of vital interest to growers. On the question of changes to the price-fixing procedures and buying, it was obvious that manufacturers saw little advantage to themselves in any change, whereas growers were convinced of the need for this. Harry Holyoake commented in 1956 that 'if any progress was to be made in the interests of growers it was virtually necessary for the manufacturers to agree'.[5] If the suspicion that there had been some manufacturer influence on the Price Tribunal were correct, it is not surprising that the board's hands appeared to be tied. Its reports were certainly inarticulate on the subject. Little wonder, too, that growers were making openly hostile and derogatory comments on the performance of the board and were doubtful of its benefit. They now knew that the minimum domestic usage could apparently be altered by board resolution. Kos Newman, also a board member, stated that the only two options open to growers were to make threats within the present system or to set up a growers' board.

There was indeed pressure on the minimum domestic usage percentage. Falling production meant less local leaf for manufacturers, and the board had reluctantly agreed to a reduction in the domestic usage requirement from 32.5 per cent (this included a voluntary component of 2.5 per cent) to 30 per cent. Manufacturers were willing to use more domestic leaf than the mandatory provision, but reduced acreage and production made this practically impossible. The situation was becoming circular. Fewer growers were producing less tobacco for less money while companies, whose output was increasing, despite the massive rise in imports, needed more local leaf to meet the domestic usage percentage. The board had no answer except to hope that its discussions with the Minister would find a way of halting the decline. The ineffectual performance of the board led many to believe that the spirit of the 1935 Act, which was intended to stabilise and regulate the industry, had been abandoned. Frustration was rising, and the chaos which had led to the introduction of the Act was nearer the surface than many had believed possible.

In 1956 the board was optimistically reporting general agreement amongst its members on the need for an incentive, in the form of increased prices, to maintain the industry. This general agreement, however, arose from pressure on the manufacturers from the board chairman, Tennent, who had criticised them for being 'unable or unwilling to produce plans to improve the industry'.[6] Representations were made to the Minister for an additional payment of sixpence per pound, over and above the price to be paid for that season.[7] Tied to this suggestion was a request for a revision of the customs duties and tariffs in order to provide greater protection for New Zealand manufacturers. Although the government instituted a tariff inquiry in March 1956,[8] no change was to be made in tariff provisions until 1958. The sixpence per pound incentive proposal sank without trace.

In 1956, for the third year in a row, the Price Tribunal failed to grant the whole of the price application. The tribunal initially informed the federation that the price for the previous year (4s per pound for flue-cured leaf) was considered adequate.[9] After strong objections and deputations from growers the price was raised by one penny per pound, still more than twopence per pound below the application. The federation was incensed and swung into attack mode once again. Letters of protest were sent to the tribunal, the Minister (now Eric Halstead) and the Tobacco Board. Despite top-level consultations, Halstead advised that the Price Tribunal could not be directed or controlled in any way. The most he could offer was 'an urgent examination of the customs and excise duties'.[10] The board was again unable to make a decision.[11]

Never content that the Price Tribunal was competent to decide tobacco prices, the federation believed that it ignored those who provided the basic information on costs and took more account of other sources. The secretary of the board, Henry Wise, partly supported growers in this. He felt that the Department of Agriculture was better equipped than Industries and Commerce to carry out surveys on farm production costs. Wise did not have a vote on the board and the chairman, Tennent, did not vote on the issue, thereby reinforcing the status quo.[12] The position of Henry Wise was increasingly irritating to the federation. Having been secretary of the board since its inception, Wise was described as a 'typical public servant'.[13] It was not this somewhat parochial observation, however, which disturbed growers. Wise's position both on the Tobacco Board and the Price Tribunal (and later on the Commission of Trade Prices and Practices which took over the functions of the tribunal), fuelled constant suspicion of his loyalties. Seen as entrenched in the status quo, he was also suspected of succumbing to the influence of manufacturers. In addition, Wise was thought to hold deep-seated dislikes of certain federation representatives, especially the previous secretary, Noel Lewis.

The federation itself was again attempting to put forward constructive suggestions for industry changes. A comprehensive plan was presented to the board and the Minister late in 1956. Backing away from a switch in departments, the federation now elaborated on the formation of a marketing authority with no manufacturer representation. The authority would consist of growers' representatives and some government nominees. It was suggested that the authority would recommend leaf prices directly to the Minister, and would control other aspects such as licensing, collection of levies and administration of the crop insurance fund. Changes to buying procedures were also planned, with an established range of grades and greater growers' representation on the buying floor. The plan included the setting up of a Tobacco Advisory Council, with full industry representation, which could advise government on policy matters. After hearing Fred Wills outline the plan at a Ward Two meeting in October, 1956, H.A. (Hec) McLaren declared his support 'for the time is right for a change'.[14]

Supporting the emergence of a viable plan, a *Nelson Evening Mail* editorial thought that, if the industry was unanimous, changes could be made.[15] But the participants were far from unanimous and the changes never came. By March 1957 the *Mail* was again regretting the existence of two grower mouthpieces. Noting that the federation plan had not found favour with the government, the *Mail* warned growers against 'automatic price increases'.[16] In the same month Halstead visited the area, meeting with both the full board and federation members. The visit convinced the Minister that the price-fixing machinery was indeed the main source of grower dissatisfaction and he undertook to examine the procedure. In the meantime he urged growers to back their representatives and advised that the current price application should go ahead as usual.[17]

The pattern was now predictable. The application for 4s 3 1/2d was lodged in early March 1957. No response was received until 18 April, when the tribunal advised prices again more than twopence below the application. Again the federation made strong submissions and again the tribunal restated its decision. This time the attack on the tribunal was immediate and direct. Freddie Wills declared that not only was the tribunal not competent to judge what growers should receive, but he added that it had been in possession of the private accounts of about a hundred growers, supposedly supplied by a manufacturer. Kos Newman expressed the belief that the policy of the Price Tribunal was to keep prices down. Its composition of two city men was inappropriate as far as the tobacco-growing industry was concerned. Other leading figures within the federation were also sounding fed-up, and not just with the price situation. There were calls from Harold Holyoake for a change in the set-up, and from Horace Dutton for more effective ways of gaining growers' participation than ward meetings. Dutton believed that federation affairs were not in order at that time and that more progress would be made if all growers were united. Ward meetings were all very well, but he felt that special general meetings should be held to discuss policy matters.[18]

Once again growers were restless over the divisions within their ranks and the possible effect this was having on their dealings with the Price Tribunal and other official bodies. Unification *was* in the wind. In July 1957 the association met to discuss its position. Claiming that the association had accomplished its main purpose of retaining the status quo, the chairman, Bill Kenyon, felt it was now time to 'settle the differences between the two growers' bodies'.[19] At this meeting a motion to disband the association, if the Tobacco Board agreed to meet any financial obligation, was lost by one vote. However, since this was the group's annual meeting and only fifteen votes were recorded, it seems that support for the association was waning and that the offer of fence-mending was made from a weak position. It also seems that the Tobacco Board had used a measure of pressure on the association. Restating its desire to see one group representing growers, the board hinted that while it could finance only one growers' group, it would consider meeting reasonable expenses if the association went out of existence.[20] Freddie Wills and the federation were unmoved by these tentative overtures. The breakaway growers would be welcome to join their fellow growers 'without discrimination or rancour',[21] but Wills stressed that the federation remained the one recognised growers' organisation. In relation to official matters this was true, although success was limited. Officialdom was fast losing patience with the long dispute and the annual round of dissension over prices.

On a visit to Wellington in 1957 Fred Wills met with the Minister of Industries and Commerce. Halstead suggested the formation of a Price Advisory Committee to consist of two growers, Henry Wise and a representative of the Department of

Agriculture. Later, Wise told Wills that he (Wise) would be chairman of this committee.[22] Obviously unhappy with this suggestion, Wills informed the Minister that the proposed committee would not solve the present difficulties within the industry. The position was again stalemated. In September 1957 the Minister wrote to the federation suggesting ominously that if the federation did not like the system under the Price Tribunal the only alternative was to decontrol the industry and let domestic leaf find its own price level on the open market.[23] Alarmed, the federation sent a delegation to Wellington to seek the support of the Minister of Agriculture, S. Smith. The delegation stressed that growers would need a guarantee of protection under any deregulation of the industry.[24] With a general election looming, Smith could make no firm commitment. The election brought the Labour Party into power, with Gerry Skinner as Minister of Agriculture and Phil Holloway as the new Minister of Industries and Commerce. Skinner promised to consult with Holloway to address the federation's concerns but little changed for growers. The Price Tribunal remained the arbiter of prices and the association remained in existence.

Conflict between the two organisations was now entrenched, especially within the federation executive. Annual reports prepared by Fred Wills, Noel Lewis and later Jeff Bradley in the early 1950s, are mainly litanies justifying the federation's stand and its efforts to remain the sole representative of growers.[25] In the second half of the decade the matter of division was seldom mentioned. More pressing matters of price, labour shortages, and the very survival of the industry occupied the minds of both the executive and the majority of growers. The division remained beneath the surface, however. Early in 1958 the association came into open conflict with the Tobacco Board, which it had previously supported almost unconditionally. Stung by a reference in the board's annual report to the association as a 'recent disruption' to the industry, Bill Kenyon retorted that the association had been in existence for five years. It had sprung from the bad relationship between the federation and the board, and its main hope was to restore harmony between growers and their controlling body. Kenyon contended that the board had failed to deal with issues within the industry. It was now trying to 'get out from under' by shifting the blame onto the split in growers' ranks. The board also appeared to be attempting to force reunification of the two groups without any change to the underlying causes of disagreement. The association was willing to negotiate but 'it would not be intimidated by the board or anyone else'.[26]

Against this unsettled political backdrop the 1957-58 season proved a particularly difficult one. Wet weather in November and December interrupted planting and favoured attacks of stem rot. Dry, hot weather in January brought considerable wilting and the cooler than average temperatures in February and March delayed the ripening of the crop. Overall production was slightly up, mainly through an increase of just over 100 acres in the area under cultivation. Grower numbers continued to fall and were now down to 411. If the weather was dismal, so were growers. The Price Tribunal again rejected their application, once more setting the price at the same level as the previous year. It seemed that growers were powerless to affect the decline in their industry. Without a miracle, the industry would continue to shrivel under the continual lack of official support.

If little changed on the political front, things were changing on tobacco farms. As the 1958-59 season got under way, growers were more interested in the range of mechanical advances available to assist with the crop and alleviate labour shortages. Several innovations of the 1950s were to dramatically change the way in which tobacco was harvested and cured. Probably the most significant of these was the development of the down-draught kiln. The principle of down-draught drying was already used by companies in redrying leaf, and in experimental work at the Tobacco Research Station. In this type of kiln, the heat for drying was provided from an external furnace chamber and forced downwards through the leaf by a large fan in the top of the kiln. With no heated pipes below the hanging leaf, the danger of fire was much reduced. As the system used heat more efficiently, a greater quantity of leaf could be cured more effectively – the capacity of the down-draught kiln was up to 20 per cent greater than the up-draught. Curing time could be closely controlled and, as the curing time was less, fuel consumption was reduced. In addition, the new kilns were usually only five racks high. This compared with eight and a half in most up-draught kilns, so that loading was easier, especially if the new movable racks were being used.[27]

In 1953 Russ Goodall of Riwaka agreed to be a guinea pig for the new kiln. A pioneer tobacco grower, Russ was sick of hand-stoking kilns with wood and coal and spending nights keeping watch on his highly flammable crop. His first down-draught kiln was 'a Heath Robinson outfit', but it was oil-fired and Russ no longer had to chop wood after tea. Growers were undoubtedly interested in the new-style kiln but many were sceptical, if not scornful. One such grower called at the kiln during drying and declared flatly 'you can't do that', to which Russ replied shortly, 'Well, we're doing it.' The visitor sought to prove his point by choosing and marking a particular leaf and inspecting it each day. When it cured more than satisfactorily, he claimed that it had been 'switched'.[28] Others took less convincing and down-draught kilns spread slowly throughout the decade. By 1957-58 there were more than a hundred down-draught kilns in use around the district.

In 1958 three growers were curing tobacco using steam boilers to provide heat. S.J. (Steve) Emerre had trialled the new system in 1957; Fred Hamilton and Kos Newman now followed suit. Installed by a local engineering firm, the combination of the down-draught kiln and steam heat had 'practically eliminated the risk of fire'.[29] The system offered ease of control, economy, and speed when cooling down after drying was complete. Thermostatic control of kilns was also being advertised, claiming 'no open flames – no smoke – no fumes'.[30] For other growers the fire hazard was not so easily dismissed. In April 1959 the Motueka fire brigade attended two kiln fires within hours. Both growers lost their kilns but damage to other buildings and stored tobacco was limited. In the outlying areas growers were less likely to convert to the new down-draught system. The often marginal soils provided insufficient yield per acre to make efficient use of the greater capacity of the new kilns – even converted up-draught kilns held more than before. Lower yields also meant less return to cover the expense of replacement or conversion. The greater acreages required to provide sufficient throughput to justify conversion would also require more workers, already a problem for those distant from the towns. More workers meant more workers' accommodation, another expensive item. Even so, it was generally agreed that the introduction of the down-draught kiln was the single most dramatic development in tobacco production since its inception.

The 1950s were, however, to see several other innovations which made life somewhat easier and tobacco harvesting more efficient than before. In 1957-58 the first mechanical tobacco harvester was demonstrated on Melvin Fry's farm at Riwaka. Imported from the United States, the machine cost over £1500. In the following season there were six of the harvesters in use and by 1959-60 nineteen farmers were using new machines, some being designed and made locally. These harvesters were double-decker models. Up to four pickers on the lower level placed leaf onto endless belts, which carried the leaf up to the second level where it was tied onto tobacco sticks in the traditional manner. The great lumbering machines proved a tourist attraction for some years – holidaymakers stopped in numbers to watch the novel sight. The machines were best suited to flat land as even moderate slopes were difficult for them to manoeuvre around.

Also in 1957-58 Fred Thomas of Dehra Doon built and used bulk trailers for the first time. Replacing the traditional bins into which tobacco had been picked for many years, the trailers made picking and transporting leaf immensely easier, especially for those still picking by hand. Positioned strategically in the paddock, the trailers removed the need to carry the heavy bins in the picking field. Pickers worked out from the centre of a specified patch of tobacco towards the trailer, where they loaded their armfuls of leaf. Once full, the long, loaded trailers could be towed directly to the kiln complex, where the leaf could be passed to the tyers straight from the trailer. Where growers could combine the use of harvesters, bulk trailers and movable racks in the lower kilns, labour numbers could be greatly reduced. In addition the grower could supervise operations more directly, a distinct advantage when dealing with inexperienced workers.[31]

Irrigation units were now commonplace, and during the 1950s were largely sprinkler systems, fed by surface pipes. These pipes required moving at approximately four-hourly intervals. Having reduced the workload of curing through oil-firing and automatic stokers, growers now found themselves spending many night hours moving the awkward pipes. Despite this, irrigation brought huge advantages to growers in the traditionally drier areas in Dovedale and the Waimeas, lifting production markedly. Tobacco varieties also affected production. Virginia Gold had been introduced in the early 1950s and now flourished in the area, producing large leaves which cured to an attractive silky gold. The tall plants gave the impression of high productivity, although it appeared that, over time, the variety was inclined to lose its 'identity'. Despite this, Virginia Gold continued as the most popular variety well into the 1960s.

Innovation also came to the grading shed, with the introduction of grading belts which carried leaf slowly past graders as they worked. These made better use of workers and simplified the task of sorting into the numerous grades. Grading did, however, remain a major winter occupation for tobacco families, local women, and seasonal workers who stayed on after harvesting – 'You knew where you was gonna be for the winter.'³² The grading shed had a culture all its own. Graders discussed the goings-on of local identities and anyone else whose activities were causing local comment. Radio serials were followed avidly: *Dr Paul* and *Portia Faces Life* were favourite radio soaps and the daily episode of *The Archers* provided entertainment as the day wound to an end. Radio was so popular in grading sheds that a decade later the federation attempted to have the local morning programme, *Telephone Time*, either suspended for the season or shifted to another time of the day.³³ Presumably workers were taking time off to phone in.

Children amused themselves in and around the sheds while their mothers worked. Small children were bedded down under grading benches or on car seats. For many local women it was the only way they could earn extra income. Child care was virtually unknown and 'taking the kids' was part of the accepted routine of tobacco work. For many, especially the tobacco wife, this routine stretched from planting to grading, October to July. Not all grading was done in the grading shed, however. One woman recalls heaps of tobacco and hanked leaf gracing her living room as they graded their crop in the winter evenings to get it finished before the close of buying.³⁴

In spite of the advances in the technology of growing and curing, the industry was making little progress on the question of changes to the administration of the industry. In 1958, for the second year in a row, there were no official visitors at the federation's AGM. Henry Wise was now acting chairman of the Tobacco Board and was either unable, unwilling or unwelcome. Neither the Minister of Agriculture, Skinner, nor the industry's own Minister, Holloway, was able to attend. The federation chairman, Fred Wills, advised the sixty growers present that the ministers were discussing the suggested changes but commented sourly that 'discussions do not mean decisions'.³⁵ The board and the Department of Industries and Commerce were again under fire. Sam O'Hara commented that the board was incapable of finding solutions and that the industry had been retarded under its present department. Fred Wills now described the board as 'completely useless'.³⁶ After its own request for an incentive payment had been declined, it was eight months before the board expressed disappointment. In November 1958 the federation reported that there had been absolutely no progress in its discussions with Holloway and that the board had not met since July. The federation was made even more aware that little had changed with the change of government when the new Minister repeated his predecessor's threat of deregulation of the industry.³⁷

As the season progressed, the price application saga was repeated. This time the Price Tribunal accepted the cost-of-production figures from the Rural Economy Division but again disputed the yield factor. It reduced the price application accordingly. The federation lodged an appeal and again whipped up support around the wards but the weariness was evident. In August the Price Tribunal declined the appeal. Frustrated, the federation decided to take no further action on price but to concentrate on developing a cost structure for the future. Growers were clearly tired of the whole business, with only forty-six present at the 1959 federation AGM. Federation membership stood at 346, but voting in the board elections had also been low.³⁸ In 1959 Henry Wise, who had been acting chairman since Tennent's resignation in 1958, was confirmed as board chairman. Wise also retained his position as secretary of the board. Federation members, especially Fred Wills, were appalled. They were critical of the fact that no prior notice of Wise's appointment had been given to the federation: 'We did not know until we read it in the paper.'[39] There was obviously little hope amongst existing federation officials for a change in the administration of the board against which they had railed for so long.

In a practical sense, the industry was experiencing a minor revival. Increased mechanisation, improved varieties and several climatically good seasons towards the end of the 1950s combined to raise production to record levels. The quantity of domestic leaf produced in 1959-60 reached 7,000,000 pounds for the first time ever. This was almost 1,500,000 pounds above the former record produced in the previous season. Increased yield per acre also played a part: the average yield rose from 1583 pounds to 1882 pounds in the same year. The cultivated area was up by 216 acres and there were nineteen more growers. In 1958 Nash's Labour government reimposed tariffs on imported cigarettes and on other tobacco products. Imports declined dramatically, with imported cigarettes falling from 1,663,301 pounds in 1957 to 43,448 pounds in 1959.[40]

Labour shortages were easing as mechanisation increased, and the labour force appeared happier with pay and conditions. Ironically, while some were still making noises about the benefits of a change of departmental control for growers, the federation strongly opposed a suggestion for tobacco workers to transfer from the control of the Agricultural Workers' Union to the umbrella of the Industrial Conciliation and Arbitration Act. This transfer would have enforced a five-day week – a particular disaster for tobacco growers as overtime would have to be paid for weekend work and any hours over forty a week. For an industry with such irregular working hours, and the bulk of the work compressed into a matter of weeks, the impact would be severe.

Some growers had other matters to worry about. In 1959 the Godfrey Phillips company cut back the number of contracts it offered. The company was said to have been over-contracting for some years but had been able to on-sell the surplus to other companies. It was now unable to continue this. The managing director, F. Littlejohn, explained that the company's market share had fallen by 20 per cent. However, he was confident that Godfrey Phillips would 'come again'.[41] The company did not really come again, but the Godfrey Phillips growers, along with the whole industry, were about to be rescued. Few could have foreseen the events of the following five years, nor guessed at the direction from which the industry's apparent miracle would come.

NOTES

[1] O'Shea, P., The Impact of the Policies of Deregulation on the Tobacco Growing Industry, unpublished paper, 1993, p.2.
[2] Harold Holyoake notes, Tobacco Growers' Federation files, p.2.
[3] Tobacco Growers' Federation minutes, 21.6.57.
[4] Ibid, 26.8.55.
[5] Ibid, 7.2.56.
[6] Ibid, 1.8.55.

7. Tobacco Board report, 1956, p.2.
8. Waugh, J.R. The Changing Distribution of Tobacco Growing in Waimea County, MA thesis in geography, 1962, p.93
9. Telegram from tribunal, 16.4.56, Tobacco Growers' Federaton files.
10. Letter from Halstead, 8.5.56, Tobacco Growers' Federaton files.
11. Letter from board secretary Wise, 29.5.56, Tobacco Growers' Federaton files.
12. *Nelson Evening Mail*, 26.5.56.
13. Interview with former grower, federation chairman and Tobacco Board member John Hurley, Pangatotara, 1.8.96.
14. *Nelson Evening Mail*, 29.10.56.
15. Ibid, 31.10.56.
16. Ibid, 6.3.57.
17. Ibid, 4.3.57.
18. Ibid, 14.6.57, 21.8.57.
19. Ibid, 5.7.57.
20. Ibid.
21. Ibid, 21.8.57.
22. Tobacco Growers' Federaton minutes, 26.7.57.
23. Ibid, 25.9.57.
24. Notes of meeting with Smith, 9.10.57, Tobacco Growers' Federaton files.
25. Tobacco Growers' Federaton annual reports, 1952-60,
26. *Nelson Evening Mail*, 15.1.58.
27. Waugh, op cit, pp.101-02.
28. Interview with Russ Goodall, 10.4.95.
29. *Nelson Evening Mail*, 2.3.58.
30. Ibid, 22.4.59.
31. Waugh, op cit, p.100.
32. Interview with Russ Goodall, 10.4.95.
33. Tobacco Growers' Federaton minutes, 3.6.69.
34. Interview with Cara Strachan, Motueka, 10.10.95.
35. Tobacco Growers' Federaton minutes, 22.8.58.
36. *Nelson Evening Mail*, 5.12.58.
37. Tobacco Growers' Federaton minutes, 6.2.59.
38. Ibid, 5.10.59.
39. Undated *Nelson Evening Mail* clipping from Wills's scrapbook, Motueka Museum.
40. Tobacco Board report, 1960, p.11.
41. *Nelson Evening Mail*, 7.8.59.

CHAPTER FOURTEEN

Miracle or Mirage?

The upheaval of the 1950s was not confined to the growing sector – companies were also to find that decade traumatic. Early in the decade major changes occurred within the National Tobacco Company. In 1951 Gerhard Husheer, now eighty-seven years old, was deposed as managing director by majority shareholders. He was replaced by a new directorate which was said to have strong links with rival company Godfrey Phillips.[1] This group had apparently lost faith in Husheer and his methods of running the company, and wished to proceed on entirely new lines. The new direction was unsuccessful. In 1954 a further directorate, favourable to Husheer, was installed, with Gerhard Husheer as a consultant and adviser to the company in its rebuilding. Gerhard's sons, Torvald and Ingolf, were to be joint managing directors of the restructured company, which promptly set about entering contracts with its growers as usual. The reorganised company looked forward to re-establishing the stability enjoyed under Gerhard Husheer's long period at the helm.[2]

Imports increased almost savagely during the mid-1950s. National Tobacco's position was perhaps not as strong as the new directorate would have liked, although a profit was recorded in 1955 after two years of substantial losses.[3] But the lack of protective tariffs and the subsequent huge increase in imported cigarettes, along with the increasing popularity of tailor-made cigarettes, were eroding National's market share. This rested mainly on cut and plug tobacco, although the company had manufactured and distributed significant quantities of cigarettes since the mid-1940s.[4] In Motueka, during the difficult period of the 1950s, W.D. & W.O. Wills bought and processed some of National Tobacco's contract leaf, mainly in an effort to assist that company's growers to remain in the industry.[5] In the previous few years growers had noticed a decline in the price offered for leaf by the National Tobacco Company and some had already switched to other companies.

In 1956 the London-based company Rothmans of Pall Mall, whose parent company was Rembrandt of South Africa, announced its intention to manufacture cigarettes in New Zealand. The company's New Zealand off-shoot, Rothmans (New Zealand) Ltd, was jointly owned by Rothmans of Pall Mall and the Butland organisation of Auckland. The Butland group had been importers and distributors of Rothmans brand cigarettes in New Zealand since 1937, with Pall Mall especially flourishing under the relaxed tariff arrangement of the 1950s.[6]

The Tobacco Board reported the presence of the new company as a positive development, hoping that it would bring increased acreages, an objective of the board since the 1950s decline had begun. The company was said to have formed a favourable opinion of the quality of New Zealand leaf, and the grower members of the board, after due consideration, decided that the increased element of competition was 'all to the good'.[7] In a bid to attract growers, Rothmans indicated it would offer prices based on the cost-of-production figures produced by the Rural Economy Division,[8] a distinctly attractive feature for growers who had battled the Price Tribunal on this point for so long. On this basis the company entered contacts with a number of growers for the 1956-57 season. The government also weighed in to assist Rothmans by waiving the statutory domestic usage percentage for three years, while the company built up a stockpile of local leaf. Any shortfall was to be made up over the following four years. The board's positive report on Rothmans' entrance on the New Zealand stage was a little misleading as all manufacturers represented on the board voted against this waiver.[9]

The situation was overtaken later in 1956 as moves to take over the National Tobacco Company came from both Rothmans and Godfrey Phillips. *The Dominion* suggested in July that Rothmans was interested in taking control of the Napier-based company. Company directors were still said to be considering the offer in November when Godfrey Phillips entered the play with a similar offer to National Tobacco's shareholders. A cash offer to the company from Godfrey Phillips had been rejected in October. Although Godfrey Phillips reported acceptance by shareholders 'all over New Zealand',[10] National Tobacco Company directors remained in favour of Rothmans' offer. In addition to a share offer similar to that of Godfrey Phillips, Rothmans proposed an alternative plan which involved a capital injection of £100,000, plus the provision of equipment and skilled personnel, both technical and in management.

Early in the new year the negotiations culminated in the announcement of a merger. In April 1957 it was reported that the London branch of Rothmans was investing a total of £1 million in the National Tobacco Company. Half of this had been used in acquiring a controlling interest and the rest would be spent on a new cigarette-making plant for the Napier factory. The new operation would use up-to-date technology and provide work for skilled and semi-skilled labour as well as offering jobs for university graduates.[11] The amalgamated company was renamed Rothmans Tobacco Company Ltd. The directors were Bruce Lindeman (chairman), P.J. Taljaard, Peter Tait, R.C. Dockery and Ingolf Husheer, all of Napier.[12] Following the merger the industry was to experience a period of incentives to grow and an expansion which verged on the unbelievable given the Depression and decline of the previous few years. By 1960-61 grower numbers had risen to 549, the highest since 1948-49, and acreage exceeded 4000 for the first time in over ten years.

Throughout 1957 and 1958 there were constant calls for an increase in the production of New Zealand leaf. In March 1958 Tobacco Board chairman Tennent estimated that up to 500 more acres of tobacco were needed, although the shortfall could be met through imports. Incentives to grow were also being considered and had been taken up with the Minister of Industries and Commerce.[13] The early move, however, came from Rothmans. In November 1958 the company announced an inducement of up to sixpence per pound to be paid on extra acreage brought into production in the next two seasons. Citing cessation of imports as a leading factor in the demand for more domestic leaf, the company hoped the incentive would give growers the opportunity to produce the extra tobacco while covering any additional costs.[14] Government measures also assisted the recovery. Cigarette imports were almost totally banned under new policies introduced in 1958 by the Labour government, and raw leaf now required import licences. Imports of cigarettes plummeted to less than 3 per cent of their previous level within a year.[15] Conditions were surely never more favourable for an expansion in domestic production of tobacco. And expand it certainly did. Grower numbers increased immediately by twenty-six in 1958-59, and acreage rose by 269. Higher yields per acre brought an overall production increase of almost a million pounds in the same season. By 1963 grower numbers, acreage and production were to reach record levels.

Having dominated the domestic market for two decades, W.D. & W.O. Wills had consolidated its Motueka operation by enlarging the leaf store in 1957. Part of this extra store was required to store the leaf it now processed and handled on behalf of Godfrey Phillips, which closed its Masterton factory at about this time.[16] With Rothmans' appearance on the domestic scene, Wills now had to keep pace with the aggressive tactics adopted by the newcomer. As part of this company rivalry, some unlikely new areas for expansion were considered. Owners of land previously thought quite unsuitable for growing tobacco were now being induced to plant the crop. Approaches were made to certain Waimea landowners to encourage them to move into tobacco growing. The idea was quickly dropped but at least one would-be grower considered legal action against Wills over what was considered to be 'breach of verbal undertaking'.[17] Even without growers from this particular area, the industry boomed as companies jockeyed for position. In 1961 tobacco was virtually 'wall to wall' in the district, with every available pocket of land turned over to the crop.

At the same time Wills was conducting experimental growing trials in the Bay of Plenty at Whakatane, Te Teko, Awakeri, Paroa and Ruatoki. The company stated it was time to determine the viability of other areas in case of devastating disease eliminating the existing growing district. This would also cover the eventuality of a scarcity of land for expansion in the Motueka-Nelson area. The Whakatane area was considered to be the nearest to Motueka in climate and soil type. The experimental tobacco was of a small leaf type and it was believed the new varieties could be more successful in the Bay of Plenty than the attempts of the 1920s and 1930s.[18] The notion of insufficient land for expansion had not yet occurred to those taking advantage of the new mood of optimism around Motueka. In 1963 grower numbers reached a record at 763, acreage surged to 5878, and production rose to a record of 9,380,752 pounds. Henry Wise's board reports were sedately ecstatic: satisfaction and optimism were reflected in virtually every section of the 1964 report. The demand for leaf was so great that local community groups used tobacco growing as a fundraising activity. Using an acre or two provided by an obliging farmer, sports and service clubs and even the pipe band spent weekends and evening hours tending their crop. The returns were significant, and funded travel, uniforms, new equipment and community projects.

Tobacco growing was now an integral part of many sectors of the community. The introduction of new varieties added to the impetus of the expansion. The high-yielding Virginia Gold had been encouraged as it was more resistant to black root rot. Certainly Virginia Gold tobacco was an attractive crop to produce. A tall plant, it produced numerous large leaves which cured into golden, silky-textured leaf. Handling and grading this leaf was a pleasure, and the perception was that such handsome leaf should be more valuable.[19] R.W. (Rob) James, superintendent of the research station, described this variety as a 'producer of good-quality cutters of low filling-power', but he added that if yield exceeded 1200 pounds per acre, 'quality [went] out the window with Virginia Gold'.[20] During the 1960s and 1970s new varieties began to replace the popular Virginia Gold. Many now recognised its inability to retain its identity and found it particularly susceptible to chilling as the season progressed. Hicks-based strains proved the most successful replacement. This variety produced smaller, narrower leaves, but had the added attraction of greater resistance to a wider range of tobacco ailments.

As the industry boomed so did the township of Motueka. Serving a population of 3000-4000 by the end of the 1950s, the retail sector boasted three appliance stores, three pharmacies, three specialist menswear stores and two furnishing outlets. There were also numerous small grocers, specialty babywear and gift stores plus a Woolworth's department store. Part of this abundance was based on the annual influx of workers. In 1960 it was stated that a shopkeeper's takings could reach £700 per week during the harvest season, compared with £50 in the off-season.[21] A further factor was the income of married women working during harvesting and grading. This extra income was often saved for larger home appliances, furnishing or home alterations, thus benefiting other commercial sectors. As the 1960s progressed, new subdivisions appeared on the east of Motueka, with the sections mainly taken up by young couples or families. Many of the new homes were made possible through the availability of a second seasonal income. Service industries benefited as well, with banks, insurance companies, accountants, legal firms and real estate companies being well represented in the small rural town. Light industry also grew alongside the development of the tobacco industry. Firms to build and service farm machinery were established, along with cartage contractors and fertiliser and chemical suppliers. Motueka was already conscious of its success. Happily servicing a range of horticulture industries, the small town was one the busiest shopping centres in the Nelson province by 1960.

By now, too, things had settled within the growers' ranks. This would stand them in good stead as yet another decade unfolded its somewhat dramatic events. The 1960s began with more fireworks amongst growers. While company manoeuvres brought a surge in grower numbers, the division between the federation and the association remained. In July 1960 association chairman Bill Kenyon found himself in hot water through a published attack on the federation. The federation threatened both Kenyon and the *Nelson Evening Mail* with defamation suits. Both were quick to react: the *Mail* printed an apology and Kenyon withdrew his remarks. Although he resented any attack on the federation, Fred Wills continued his own barrage against the Tobacco Board. This intensified in 1960 when Henry Wise was confirmed as board chairman. Wills said he had no intention of coming into conflict with Wise, but told growers 'you have all read how we were taken to task'.[22] Wise had publicly commented that the levy on growers to fund the operations of the board, which many complained was an unwarranted imposition, did not cover the costs incurred by the board on growers' behalf. Salt was rubbed in the wound by his further assertion that the cost of the levy was recoverable in the federation's annual price application. This implied that the levy was ultimately paid by the smoking public in the price of tobacco products.[23]

Fred Wills retaliated by presenting a comprehensive plan for the reorganisation of the industry. This plan, if put into effect, would do away with Wise's position. The grower members of the board had wasted no time in meeting with the new Minister of Industries and Commerce, J.R. (Jack) Marshall, following the election of a National government in the 1960 general election. Marshall asked the delegation to submit a plan for the industry. This was prepared without delay and was now ready for presentation to growers. Over 100 growers met in late November 1960 to hear the details of the new plan. They first heard Fred Wills spell out a long list of the board's deficiencies. This included the 'obvious attitude' of the manufacturers and the advantage to companies of the present structure of the board. The equal representation of growers and manufacturers led to deadlocks. The situation was not helped by the addition of a government appointee who was 'obviously unwilling to be instrumental in the creation of any major changes in the industry'.[24] Grower members also objected to the ability of manufacturers to discuss growers' affairs. The same did not apply in respect to company affairs. With these limitations, the board was accused of being unable to deal with vital matters. The cost of administration was a further problem. Fred Wills pressed his case further by attacking the present system of financing the industry and of buying tobacco. It was a manufacturers' industry and their power on the buying floor was 'absolute'. The federation had little power within the industry, said Wills. Its chief function was to allow growers to discuss their concerns collectively and make recommendations to the board 'where nothing happened'.

The plan for reorganisation was then outlined. Its main focus was a complete shifting of the balance of power in the administration of the industry. Two major bodies were proposed. The first, a marketing authority, based in Motueka, with wide powers, including the licensing of growers, the collection of fees and levies, and control of the crop insurance scheme. Routine functions included the general representation of growers' interests and the employment of staff to carry out the authority's duties. The plan also supported greater use of the Department of Agriculture Act 1953, which empowered the department to encourage and promote tobacco growing. This was possibly an attempt to shift departmental control in a roundabout way. Secondly, the plan proposed a Tobacco Advisory Council, comprising growers, manufacturers and the government. This would advise the government on policy matters and oversee the industry in general. The contract system of growing would be retained (presumably manufacturers would be required to agree to this) and price control would be a matter for the Cabinet to decide. The meeting, which was open to all growers and not confined to federation members, debated the matter with interest. Fred Hamilton wanted the plan to be circulated to all growers and Hec McLaren moved to have ward meetings consider the matter. But most growers were impatient for action and a large majority supported immediate discussion of the proposals with government. Whether the proposals were ever presented to the Minister is unclear and the plan appears to have sunk without trace. Once again, there was no change in the structure of the industry.

As the season progressed, attention turned to the old perennial of labour shortages and squabbles with the Workers' Union. As early as December 1960 it appeared that labour would be short for the coming season. H.J. Allen, speaking for the Workers' Union, claimed that growers were responsible. He accused tobacco employers of being parsimonious and said wages and conditions in the industry were the lowest in New Zealand. The federation had offered a three-farthings-per-hour increase in recent negotiations.[25] Union discontent continued to simmer: a correspondent to the *Nelson Evening Mail* claimed the increase offered by the federation was 'nothing short of miserable'.[26] In January, as harvesting began, labour was still said to be short but the position was not desperate.

The 1960-61 season was slightly less favourable than the previous year, which had seen record production of over seven million pounds of leaf. Expansion had continued, however, with a further eighty-one growers and the area in production was up by 401 acres. Local builders were stretched to the limit as forty new kilns were erected to cater for this expansion.[27] Mechanisation continued, with the introduction of the 'String-O-Matic', a harvesting machine incorporating a tying machine on its upper level, again requiring less labour. The tying mechanism worked in the manner of a sewing machine, with leaf being stitched mechanically onto tobacco sticks. Very quickly, similar machines were being manufactured locally. Bulk curing of tobacco was also being trialled. For this type of curing, the leaf was held in clamps which held the equivalent of up to six of the traditionally tied tobacco sticks. This process, with its enhanced capacity, could be carried out in smaller kilns as two racks of clamps could hold the equivalent of six racks of sticks. But even with these innovations, which in any case could often be afforded by only the larger growers, labour shortage was still a problem. In mid-March 1961 workers were again 'urgently needed'. With the threat of an early frost ever nearer, the community was urged to assist by offering weekend labour.[28] Growers were said to be 'struggling through' and, with some now finishing their harvest, workers would be available to assist other farmers. Despite growers' problems during the season, the wage dispute was not resolved. In May the matter was referred to the Arbitration Court, with the union applying for a fourpence-three-farthing increase.[29] The argument was still rumbling on in November and echoes were heard into the 1962 harvest season.

The federation and the Tobacco Board finally succeeded in one endeavour which had occupied their attention for some years: the establishment of a full-time seasonal labour office in Motueka. In 1962 the Department of Labour agreed to the proposal and offered to fund the operation up to £400, with the industry to meet the balance. The office opened for business on 7 January 1963, with M.M. (Snow) Glynan in charge. When the new arrangement ceased on 26 April it was described as an unqualified success.[30] Over the next few years the scheme was continued and expanded, opening in November and carrying on into May as the grading season got under way. Snow Glynan extended the recruitment activities from newspaper, radio and screen advertising, to personal visits to likely areas. Glynan visited labour offices and Maori settlements, especially in the Bay of Plenty, Hawke's Bay and Gisborne. Glynan's activities during the season included meeting workers as they arrived in buses and arranging for them to get to their destination. Company field officers found themselves 'roped in' for some of these 'deliveries' of workers. Using their own vehicles, the field officers had to navigate the narrow gravel country roads, often in the dark, as the buses generally arrived in the evenings. At times the workers arrived, with the field officer, only to find the grower had already secured workers and no longer required them. Off the hapless field officer had to traipse, with his bewildered cargo, to find an alternative destination for them.[31] This problem also worked in reverse. Rona Hurley told of cases where workers allocated to specific growers had been 'intercepted and taken by others'.[32] Despite the difficulties, the existence of a full-time labour placement office removed many of the frustrations of labour recruitment of previous years.

Other problems associated with seasonal workers remained. The community was regularly either entertained or offended by reports of workers' behaviour. In 1964 the publication of a confidential report by the Maori welfare officer, Kia Riwai, saw local residents defending themselves against accusations of partner-swapping and generally immoral activities. The suggestion that the statements referred to locals was rejected with righteous indignation. A local taxi driver, A. G. McGregor, supported this stance, but added fuel to the fire by stating that he and his fellow taxi drivers regularly delivered both male and female workers back to their baches, from other baches, in the early hours of weekend mornings.[33] Confrontations were not restricted to amorous adventures. Groups of workers often clashed at local hotels and parties, with factional fights between Australians and Maori making the headlines during one season. Police Sergeant E. Daken pointed out that, while the majority of workers were 'hard-working, law-abiding citizens',[34] the hundred-odd troublemakers could cause considerable havoc. After several weekends of fights between the Australian visitors and Maori workers, a brawl at a Riwaka bach resolved the matter with 'the overseas contingent receiving very much the worst of the fray'.[35] A 'somewhat battered party' of Australians left the district on the first bus out of Motueka the following Monday morning.

Other disputes also distracted the industry. The friction between the federation and manufacturers continued, as tobacco companies announced the formation of the Tobacco Companies Trust. Set up by companies to assist growers who suffered crop loss not covered by the crop insurance scheme or who were in financial distress, the proposal was viewed with suspicion by some in the growing sector. The trust was chaired by the mayor of Motueka, Major H.H. Thomason, with Rowly Talbot and Wally Eginton representing growers. Companies were represented by Allan Duff from W.D. & W.O. Wills and T.A. Kyle from Rothmans.[36] The *Nelson Evening Mail* described the trust as an 'act of philanthropy', even if it was partly self-interest on the part of companies.[37] Opinions amongst growers were less complimentary. At the federation's 1961 AGM the matter was hotly debated and a motion of appreciation to companies for setting up the trust was defeated. Fred Wills claimed that grower members of the board had long been pressing for a widening of the crop insurance regulations to cover the proposed aims of the trust. Companies knew the views of the federation on the matter and should have initiated discussions at board level. G.W.B. (Gilbert) Hunt disagreed strongly with Wills. The gesture implied in the trust 'should not be flung back in the faces of the manufacturers'. Even stronger were the views of M.A. (Maurice) Cederman, who described the trust as 'cunning tactics and propaganda by the manufacturers'.[38] Later Cederman was publicly criticised for his attack on manufacturers. A *Nelson Evening Mail* correspondent thought that growers wanting manufacturers to consult them before such moves was 'too ridiculous. The federation doesn't even consult growers before they take a course of action.'[39]

Rumbles continued following the AGM. Rowly Talbot attempted to gain support for changes to the federation's structure, especially to the ward system and the composition of the executive. Talbot wanted the executive and the federation chairman to be elected by the full membership at the AGM each year. This move was designed to break the control of the executive, which was made possible by the election of a chairman from within its own ranks. Talbot described the present method of representation on the executive as undemocratic. Wards containing large numbers of growers had only the same numerical representation as smaller wards. Talbot had stayed within the federation, but he obviously held similar views to the supporters of the rival association. Naturally, Fred Wills opposed Talbot's suggestions. He reiterated that the system had been designed to prevent the domination of small wards by larger areas.[40] The association also constantly irritated Fred Wills by negotiating separately with the Workers' Union. Wills maintained that the federation was the only body with the right to enter into such negotiations. The association, on the other hand, claimed that the federation was holding out in union talks simply because of its antagonism to the association.[41] The industry had changed in many ways, including technical changes and the rapid increase in grower numbers, but it seemed that the old adversarial attitudes and personality-driven tactics remained.

Many of the new growers of the 1960s, however, were not interested in the animosities of the past. In 1962 it was clear that a change was coming. Fred Wills was defeated in the annual ballot for grower members of the Tobacco Board, with Kos Newman and Wally Eginton being the growers' choice. At a somewhat sober meeting following the 1962 federation AGM, Newman was elected chairman, ending a twelve-year reign for Freddie Wills. By January of 1963 talks between the erstwhile enemy groups had taken place and amalgamation was under way. At the final meeting of the association in April a resolution to wind up the organisation was carried. The civil war within the tobacco-growing industry was ended with barely a backward look. The popular Kos Newman headed the united body which now represented all growers, many of whom were new to the industry or young second-generation growers.

Kos Newman represented an intermediate generation of growers. Not one of the pioneers, he had taken up land in the Motueka Valley, under the Rehabilitation Scheme, where he developed a large acreage of tobacco. Having served with the air force in India and Burma, where he earned a DFC and Bar, Newman took an active interest in the welfare of returned servicemen. Elected to the Tobacco Board in 1948, Kos Newman was described as courteous, cheerful and popular.[42] His chairmanship spanned the divide not only between the association and the federation, but between old-timers and the numbers of young growers. These growers brought a fresh approach to the industry. Through the added influence of young farmers' clubs and Jaycees, many of the young growers were willing to speak out and to participate in the federation's affairs. This approach became apparent with the emergence of such young growers as John Hurley, Gerald Hunt, Peter Heath and, later, Graeme Emerre. All except Heath were second-generation growers.[43] Newman's ability to relate to all sectors of the industry promised to do much to advance the cause of tobacco growers. The local manager of W.D. & W.O. Wills, Peter Wild, described Newman as 'a real diplomat',[44] an obvious tribute to his personal skills.

The secretaryship of the federation was also to experience changes in the 1960s. When Jeff Bradley left the district in 1961 the federation was in urgent need of a replacement. The position was filled by K.C. (Ken) Collins, the retiring headmaster at Motueka South School. In 1964 the federation decided to appoint a full-time secretary-accountant. The executive believed that the complexity of price applications required a suitably qualified person and that, ideally, the federation should employ its own accountant. N.C.L. (Norman) Holloway of Blenheim was appointed and took over from Ken Collins in September 1964. Holloway lasted less than three months. The British accountant appeared rather patronising, and he assumed a familiarity which federation leaders found uncomfortable. In December, dissatisfied with various aspects of his tenure, the executive moved to dismiss him. More than a year of wrangling ensued. Holloway threatened legal action, although it never ensued. He hinted, somewhat waspishly, that the federation's annual price applications were unsupported by 'local accountants or ... certain members of the executive',[45] implying that the applications were, in fact, a fabrication. The episode proved extremely unpleasant and the executive was badly burned. Ken Collins was persuaded to fill the position for the time being.

The industry suffered two major losses in 1964 and 1965. In June 1964 Steve Emerre, a long-time grower, association supporter and Tobacco Board member, was killed in a road accident. Emerre was a cheerful, outgoing personality with a record of involvement in a number of community organisations. Like many other growers he was also a keen and talented sportsman. His death, at the age of fifty-four, was seen as a great loss. Emerre's place on the Tobacco Board was filled by a familiar face, that of Dick Stevens. With the buy-out of St James by W.D. & W.O. Wills in December 1963,[46] Stevens was now free to represent growers without any suggestion of divided loyalties. He was to prove a commanding presence in the industry for the next fifteen years.

Another shock for the federation was the sudden death in May 1965, of the current chairman, Kos Newman, also in a motor accident and coincidentally on the same bridge as the accident which had killed Steve Emerre. Having been instrumental in the reunification of the industry, Newman had been a valuable element in maintaining stability within the federation. With some foresight, he had already warned growers to beware of dependence on company finance. Newman likened this to another form of the welfare state, especially where such finance included living allowances. He also believed the boom years were actually bad for the industry. Newman wisely saw that it was the government of the day which 'could make or break the industry'[47] through

W.C. (Fred) Wills.
Controversial chairman of the Tobacco
Growers' federation 1949–1963.

Rona Hurley, long-time grower, crop
damage assessor and a buyer for St James.

W. (Bill) Kenyon, president of the
Motueka Tobacco Growers' Association,
which opposed Fred Wills' management of
the Federation.

Hand-picking tobacco remained common
on many farms, especially for the lower
leaves (sand leaves and lugs).

Grading and hanking were regular sources of income for women during the winter.

A Grading Day

The morning is both dark and cold,
Time to rise, I've just been told;
How I would love just one more hour,
And time to waste after a shower!
But I have a date with my favourite weed,
Without a smoke, would be hard indeed.
So I must wend my frozen way
And do my stint in grades today.
Once in the shed, my coat I doff,
And feel my hands are coming off!
My fingers, worked by powers remote,
Cling desperately to my coat.
They just refuse to do my will,
They've lost their natural knack and skill.
I warm them up and work begins,
And by lunchtime we start to sing,
By then we've listened to Dr. Paul,
And by his actions been enthralled,
Then Portia bravely faces life
In this hard world of toil and strife.
Will she lose Chris or will she not,
Surely he won't turn from her, the clot?
A voice breaks in, with dust quite gruff:
I look around in quite a fuss,
'And now I'll take the little roughs',
Surely, he can't be meaning us?
'I'll take the little brights as well!'
I push them out and start to yell.
My ankle knocked against a stool,
I really feel a clumsy fool.
Hip hip hooray, here comes the tea,
The day is saved for little me.
Later we tie leaf into hanks,
My stool stays put now, many thanks!
I hank and dream and dream and hank,
Odd odes I make, both good and rank.
And when at last the day is done,
Behind the hills has sunk the sun,
Back to the cottage, dinner to get,
Spuds cooked, meat braised, table set.
Time to relax inside the shower,
I start to blossom like a flower.
Over to the neighbour's to see TV.
How nice they are to Heather and me.
Back to bed, our hair we lacquer,
And then we dream about 'tobacca'.

The atmosphere of the grading shed is neatly captured in words.

In the 1950s gangs were still large. This gang at Hickmott's Riwaka farm enjoys a 'break-up' celebration at the end of harvest.

Noel Lewis, Motueka solicitor and secretary of the Tobacco Growers' federation from 1938 to 1953.

Sedrick F. Brame, acknowledged tobacco expert. Brame came to Motueka in 1928 and returned in the 1940s to work for National Tobacco Co.

The new complex built by the National Tobacco Company, on the corner of High St and King Edward Street, Motueka, was taken over by Rothmans Tobacco Co. in 1957.

At Riwaka, the St James Tobacco Co. was based on Dick Stevens' farm.

K.J. (Kos) Newman, became chairman of the Federation on the resignation of Fred Wills and set about reuniting the industry.

W.J. Eginton, long-time mayor of Motueka, continued the unification after Newman's death in 1965.

An aerial view of the Tobacco Research Station at Brooklyn, in the 1960's.

Mechanisation made great strides in the 1960s. Pictured is one of the first harvesting and tying machines to come into the district.

The tying machine was the next major mechanical innovation, reducing tying staff dramatically.

Hundreds of workers were still required for the annual harvest. M.M. (Snow) Glynan, labour placement officer, greets workers arriving on the evening bus.

The Federation carried out extensive advertising to obtain sufficient workers. Cinemas all over the country carried these seductive images.

Trucks loaded with crated tobacco arrive at the Nelson wharf c1964. W.D. and H.O. Wills exported several hundred thousand pounds of leaf in an attempt to reduce their stockpile of domestic tobacco.

A Dovedale gang takes a break – note the long trailer now used for carting leaf from the field.

Growers listen as company advisor, A.L. (Bert) Black explains the grade requirements for the coming buying season c1966.

Growers attending a Research Station Field Day c1967.

R.W.S. (Dick) Stevens, a commanding figure in the tobacco industry, member of the Tobacco Board for more than twenty years and chairman of the Federation 1966–1976.

The down-draught kiln has come to stay. Here, on the left, the new kiln contrasts with the old.

Wet weather did not always mean 'stop work'.

R. (Bert) Thompson, retiring Superintendent of the Tobacco Research Station, greets his successor, R.W. (Rob) James.

A plant affected by Verticillium Wilt, a major tobacco disease which occupied the researchers into the 1980s.

Council of war. Federation members meet with Q.C. Frank O'Flynn to discuss progress at the Committee of Inquiry.

George Laurence, Committee of One for the Committee of Inquiry in the Tobacco growing industry 1966–1969. Laurence became a life-time supporter of growers.

E. (Ernie) Hardaker, proud groundsman at the Rothmans factory. Ernie's lawns and gardens were an attractive surround to the landmark clocktower.

Ward committees were an integral part of the political structure of the industry. Pictured are Ward One committee members in 1971.

Tobacco plants completely devastated by a hailstorm. Damage sustained from frost, hail and flood was covered the crop insurance scheme introduced in 1945.

Keith Holyoake, now Prime Minister, visits the buying floor at a local factory.

The purchased leaf proceeds through the Proctor machine – the hanked tobacco is placed over sticks once more to be re-dried.

supportive or adverse legislation. Newman's place on the Tobacco Board was taken by one of the new, younger growers, John Hurley. A son of Rona and Dan Hurley, he was to prove another determined advocate for his industry until the 1980s. The federation executive elected Wally Eginton to take over the chairmanship, a clear sign that growers had largely resolved their former differences. The speed at which the industry was expanding left little for growers to be divided over.

The federation took another step in establishing itself as a centralised growers' organisation with the publication of the first *Tobacco Growers' Journal* in January 1964. A committee under the guidance of Rowly Talbot coordinated the new venture, which was well received by growers. Produced on good-quality paper and supported by industry advertisers, the journal was published six times a year. It contained articles of technical interest, comment from industry participants, biographical features and items of general interest and entertainment. The journal was distributed free to all growers; any shortfall between advertising revenue and costs of production was met by the Tobacco Board. The inaugural issue of the journal was well supported, with contributions from Keith Holyoake, now Prime Minister, Jack Marshall, Minister of Industries and Commerce, and Henry Wise. Congratulatory messages from all three manufacturers were also included. W.D. & W.O. Wills was especially effusive in its support, being 'proud to boast the longest association with growers in this country'.[48] This, of course, was arguable. Rothmans, through its predecessor the National Tobacco Company, claimed a longer relationship with New Zealand growers in the same issue of the journal.

Gathering material, advertisements and other articles proved a time-consuming task for successive federation secretaries, assisted by the journal committee. The journal collection, covering the years from 1964 to 1980, retains the original quality of presentation and content. Although issues were restricted in later years, each is of interest in itself, and collectively they provide a reflection of the fortunes of the industry through the 1960s and 1970s. In later years this reflection is less positive, but in 1964 the journal mirrored the buoyancy of an industry on the crest of a wave of development which seemed set to continue for the foreseeable future.

But all was not well beneath the surface of the industry. In 1965 manufacturing companies, W.D. & W.O. Wills in particular, found themselves with alarming quantities of domestic leaf in stock. The battle to monopolise growers had resulted in companies contracting for, and purchasing, far more leaf than they could realistically use. Bringing in greater areas of marginal land had resulted in large quantities of low-quality tobacco to which companies were committed under the 'run of the crop' buying system. In 1965 the total company stockpile amounted to more than nineteen million pounds of leaf, compared with ten million pounds in 1959.[49] In an effort to eliminate its stockpile of New Zealand leaf, Wills was now using more than fifty per cent local leaf in its products. The end result was said to be suffering from a lack of balance in the blending.[50] In addition, Wills packaged local tobacco for export in both 1963 and 1964. Hailed as an opportunity for New Zealand growers, it is more probable that the intention was to relieve pressure on the huge stockpile. The shipments were impressive when photographed proceeding in a convoy of trucks along Rocks Road in Nelson, but the quantities were insignificant in relation to the millions of pounds being held by companies. Some 220,000 pounds were shipped in 1963, followed by 100,000 pounds in 1964 – the stockpile was virtually undented.

Rothmans' usage was somewhat less and it was able to ride the rising tide of domestic production while increasing its market share at Wills' expense. A factor seldom mentioned, but evident in board reports, was the growing significance of imports of raw leaf from African countries, especially Rhodesia (now Zimbabwe), Malawi and South Africa itself. In 1957, 26,937 pounds of Rhodesian leaf came into New Zealand. By 1963 this had risen to 1,314,770 pounds.[51] Rhodesian tobacco could be imported very cheaply and was said to be similar to New Zealand leaf in its usage qualities. Imports of the more expensive American leaf, which had historically made up the bulk of unmanufactured tobacco imported into New Zealand, had fallen by three million pounds since 1955.

Drastic measures were needed to rescue the situation. In July 1965, just as growers were expecting to renew their annual contracts, Wills announced a 20 per cent reduction in contracted acreage, a decrease of approximately 500 acres. Growers were stunned. After several years of almost unrestrained, company-induced expansion, they had been lulled into a sense of invulnerability. Many could scarcely believe what was happening. Once again the confidence of the industry was shattered. Urgent negotiations began between the federation, the board and the manufacturers. An air of recrimination pervaded these discussions, as companies had given an undertaking the previous November that they would maintain current acreage levels. Even Henry Wise took a critical stance. Board reports had, for several years, been largely self-congratulatory on the expansion in the industry. Now, Wise openly blamed the 'scramble for acreage' for the situation. In seeking to corner the market, companies had overstepped the mark and now sought to recover by placing the whole burden on growers.[52] Wise also claimed that the companies' November undertaking had later been confirmed in writing. Henry Wise did not escape his share of blame, however. Growers believed that Wise had known the stock position of companies since 1959 and should have realised that the industry was heading for trouble.[53]

Under pressure, the companies came up with a set of suggestions for the stabilisation of the industry. These were to limit yield to 1600 pounds per acre and place a hold on the number of contracts. This amounted to control through yield, rather than acreage. Companies also agreed to 'open and frank' discussions on the future of the industry.[54] Growers were unimpressed by these proposals, viewing them as just another way of placing the onus on growers to fix manufacturers' own problems. Wally Eginton told growers that companies had induced and persuaded growers into the industry in any way possible. When growers had become concerned about stock levels they had been told not to worry. Few concessions were made to growers to help them through the crisis. Growers struggling with loan repayments to companies had been told that reductions in loan repayments would not be permitted.[55] None of the recriminations or accusations deflected the intent of Wills to reduce its contracted poundage. Hasty arrangements were made for the bulk of Wills' affected growers to be transferred to Rothmans. Later, Wills was persuaded to pay £50 per acre compensation to its growers for lost income, a modest offer in view of the estimate that the reduction in total income from tobacco for the season would be £220,000. The situation was ambiguous, however. At least one grower received a letter from Wills requesting him not to destroy any surplus marketable leaf and some of this leaf was later purchased by the company.[56] But the industry was seriously wounded. In December 1965 there were rumours that there would be a further 10 per cent cut the following year.[57]

Convinced of the validity of its case, the federation, especially Dick Stevens, exerted constant pressure on the government. In 1966 it was announced that an inquiry would be held into the industry. A committee of one, George Laurence, was appointed with the brief of inquiring into 'the economic position of the tobacco-growing industry in New Zealand'.[58] Laurence had been the government appointee on the Tobacco Board for some years and had also served on the Price Tribunal. He was, therefore, relatively familiar with the structure and workings of the tobacco industry. The committee was required to report back to the Minister within three months. Growers pinned their hopes on the outcome of Laurence's deliberations and looked forward to a positive report before the end of the 1966-67 season.

Their hopes were ill-founded. The inquiry wound on until 1968, and did not present its final report until September 1969. The proceedings of the inquiry were largely dominated by the factional divisions within the industry. This time animosity re-surfaced between growers and companies. Bitter battle lines were drawn and opposing lawyers fired verbal and written bullets made from years of underlying tension and suppressed conflict. To those who believed their industry had matured and stabilised over its forty-year history, the proceedings of the inquiry were to prove something of a shock.

NOTES
[1] Interview with John Husheer, Napier,13.8.95.
[2] *Daily Telegraph*, Napier,. 26.5.54.
[3] *The Dominion*, 8.11.56.
[4] T. Husheer, notes.
[5] Interview with Bert Black, 21.6.96.
[6] Interview with John Husheer, 13.8.95.
[7] *Nelson Evening Mail*, 26.5.56.
[8] Ibid.
[9] Ibid, 26.5.56.
[10] *The Dominion*, 19.11.56, 30.11.56.
[11] *Nelson Evening Mail*, 1.4.57.
[12] Ibid, 14.6.57.
[13] Ibid, 26.3.58.
[14] Ibid, 20.11.58.
[15] Tobacco Board report, 1961, p.12.
[16] Interview with Bert Black, 21.6.96, and draft of Black's evidence for case against Wills, c. 1991.
[17] Interview with Bert Black, 21.6.96.
[18] *Nelson Evening Mail*, 30.11.60.
[19] Interview with John Hurley, 1.8.96.

20 Notes of an address to the Tobacco Growers' Federation executive, federation files, 22.2.67.
21 *Nelson Evening Mail*, 27.8.60.
22 Ibid, 29.8.60.
23 Ibid, 16.7.60.
24 Ibid, 22.11.60.
25 Ibid, 23.12.60.
26 Ibid, 8.2.61.
27 Tobacco Board report, 1961.
28 *Nelson Evening Mail*, 17.3.61.
29 Ibid, 13.6.61.
30 Tobacco Board report, 1963.
31 Interview with Bert Black, 11.5.95.
32 *Nelson Evening Mail* clipping, c. 1958, personal papers of Rona Hurley.
33 *Nelson Evening Mail*, 11.2.64, 12.2.64.
34 *Nelson Evening Mail* clipping, c. 1958, personal papers of Rona Hurley.
35 Ibid.
36 *Nelson Evening Mail*, 2.7.61.
37 Ibid, 12.7.61.
38 Ibid, 28.8.61.
39 Ibid, 4.9.61.
40 Ibid, 28.8.61.
41 Ibid, 8.9.61.
42 Tributes to Kos Newman, *New Zealand Tobacco Growers' Journal*, July 1965. *Nelson Evening Mail*, undated clipping from Wills scrapbook, Motueka Museum.
43 Peter Heath, however, married into a Dovedale tobacco-growing family.
44 *Nelson Evening Mail*, undated (but c. May 1965) clipping from Wills scrapbook, Motueka Museum.
45 Letter from Holloway to Tobacco Growers' Federation, 17.8.65 The correspondence relating to this episode makes fascinating reading. The federation was impressed with Holloway's account of his previous experience but found themselves unable to relate to him as an employee. Federation files.
46 Tobacco Board report, 1964.
47 *NewZealand Tobacco Growers' Journal*, January 1964, and *Nelson Evening Mail*, 24.7.64 and 16.9.64.
48 *New Zealand Tobacco Growers' Journal*, January 1965, p.15.
49 Inquiry notes of F.D. O'Flynn and Tobacco Growers' Federation submission to the Committee of Inquiry –'Stabilisation of the Industry', p.2. Federation files.
50 Interview with Bert Black, 21.6.96.
51 Tobacco Board report, 1964.
52 *Nelson Evening Mail*, 8.12.68.
53 Ibid, 21.1.66.
54 Ibid, 18.12.65.
55 Ibid, 21.12.65.
56 Letters from Wills to J. Hurley, 20.12.65, 4.3.66, federation files.
57 *Nelson Evening Mail*, 9.12.65.
58 *The Economic Position of the Tobacco Growing Industry in New Zealand*, Report of the Committee of Inquiry, Government Printer,1971.

CHAPTER FIFTEEN

Gloves Off

The aftershock of the Wills' contract cuts saw grower numbers fall immediately, from 728 in 1964-65 to 585 the following season. Part of this reduction was the ending of contracts for sports and service clubs and other fundraising ventures. In addition those growing very small amounts found their reduced contracts hardly worth the bother. While total acreage fell only marginally, production reduced dramatically by three million pounds. This was partly through climatic factors and partly through the contract cuts, plus the imposition of tighter planting and buying standards by companies. Growers claimed that leaf which would previously have been purchased by manufacturers was now rejected. Some had more than 20 per cent of their crop classed as unacceptable. One grower had so much rejected from a load that he had 'burnt it in a fury'.[1] In addition, yield per acre was down significantly on the previous year,[2] reducing growers' returns still further.

In 1966-67 grower numbers were down to 529 and by the end of the 1960s numbers were at their lowest level since 1958. The boom was truly over. In his report to federation members in 1967, chairman Dick Stevens claimed '[t]he growing industry is on its knees and many growers are insolvent'.[3] This was illustrated by growers' claims in the press. Living on meagre weekly allowances, some felt they were nothing but peasants. One asserted he was receiving less than half the hourly rate of his workers. Balancing on the edge of financial ruin, such growers were still planting and hoping, having few other options. Selling up would not bring enough to pay off debts. A few growers were working off-farm in order to live, but this did not cover debts incurred in tobacco growing. 'I suppose I could go bankrupt,' said one, 'but what would that solve?' This grower claimed that everything he had 'went down the hole ... I just became a twentieth-century serf'.[4] Approached by the writer of this study, manufacturing companies felt unable to comment pending the outcome of the Committee of Inquiry.

Some claimed the inquiry was just 'a smoke screen ... until the elections in November'.[5] Others saw it as an opportunity to revisit the structure of the whole industry and re-submit plans for the stabilisation of production and usage. Whatever growers thought, the inquiry went ahead. At the first sitting George Laurence made the point that the inquiry was not a court and he hoped the proceedings would be as informal as possible. The prime objective of the hearings would be the 'acquisition of full and reliable information and ... the soundest interpretation of its significance. (I)n these circumstances, formality should not claim highest priority.'[6]

Laurence announced that the major points to be examined by the inquiry included:

- How well equipped the industry was to keep up with technological advances;
- Whether farmers' attitudes or financial restrictions were impeding the adoption of new methods;
- Whether a widening of the varieties grown would benefit the industry;
- What could be done to improve confidence and farm management skills;
- At what point should manufacturers substitute domestic leaf for imported leaf and for what reason should this be done.[7]

All interested parties would be encouraged to present submissions, and Laurence hoped the inquiry would 'carry on rather as a discussion group observing the proper courtesies'.[8] Any person or organisation who felt their information was confidential or that its publication might be embarrassing would be given the opportunity to make private submissions to the committee.

Laurence had hopes of an informal and friendly process. It became clear, however, that the major participants viewed the inquiry as more of an industry showdown. They chose their weapons accordingly. The Tobacco Growers' Federation employed the services of F.D. (Frank) O'Flynn, Queen's Counsel of Wellington, who was assisted by Frank Bryson, a Motueka solicitor. O'Flynn had impressed the federation with his role in the Manapouri case against the government, and a successful result in a local case. He quickly grasped the details of the industry and developed a sympathy for growers. Frank O'Flynn brought a new dimension to the presentation of the growers' case; previously content to be represented by federation officials, growers now found themselves used dramatically in the courtroom manner of the Queen's Counsel.

The manufacturers' representative, the doughty local solicitor Noel Lewis, was quick to take up O'Flynn's challenge. With his many years as federation secretary, Lewis's insight into both the industry and the workings of the growers' organisation was no doubt invaluable to the companies. In addition, Lewis probably held few fond recollections of the current federation chairman, Stevens, who had played a leading role in the association, which had forced Lewis's resignation as federation secretary. For his part, Dick Stevens pulled no punches in his assessment of Lewis. Recalling that Lewis had been political organiser for Keith Holyoake in the 1930s, Stevens believed that his attacks on the Tobacco Board in the 1950s had been an attempt to 'pay off old scores and to advance the claim for political patronage'.[9] Obviously Stevens expected no quarter from the former federation secretary. With the scene thus set there were to be some acid exchanges, with only the urbane influence of Laurence to calm the proceedings. George Laurence took a painstaking approach to his job as the committee. He examined every detail of the industry available to him and sought to understand as fully as possible every facet which might affect the current position of the growers.

Little animosity was evident in the early stages of the inquiry as Laurence set out the way in which he wished to proceed. Questionnaires had been sent to all growers in order to provide a comprehensive picture of the position, both financial and philosophical, of the tobacco-growing sector. A number of growers required persuasion to return the questionnaires, which contained personal as well as financial information. Not all were convinced of the confidentiality guaranteed by Laurence. Analysing the information from these questionnaires was to prove difficult and contentious. The federation maintained its position that the figures proved growers were under severe stress. Manufacturers took the familiar line of the Price Tribunal in being suspicious of information produced by those who had called for the inquiry.

Attacking the questionnaire information, Lewis said 'manufacturers regard the information ... with distrust for more than one reason'.[10] Besides being provided by those who had called for the inquiry, the information covered only one year, which had been the worst season since 1935. The accounts, in fact, covered three years, but the particularly bad results of the 1965-66 season were precisely why growers had called for an inquiry. Lewis pressed for the names of growers to be given so that companies could verify the figures from their own records. George Laurence was stung at this point to comment that he 'would be more impressed... if the manufacturers... were to produce data to show how wrong we are...'.[11] Reprimanding Lewis for casting doubts on the sincerity of growers, Laurence was also annoyed that companies could be holding information which would be helpful.

The participants returned to this point later in the inquiry when Lewis again requested the names of growers. While Laurence felt he could not divulge specific information he did not feel it would be breaching confidence to reveal which growers had cooperated. O'Flynn disagreed strongly, saying that if companies mistrusted the information itself, the federation was now somewhat suspicious of why companies wished to know which growers had provided it. The issue was not resolved at the public hearings, Lewis's suggestion being that Laurence should confer with manufacturers in Wellington. When O'Flynn asked the reason for this, Lewis responded, ' I think you had better get that in Wellington.'[12] With such cryptic statements it is small wonder that Laurence was at times frustrated, and the federation often suspicious of the way in which the companies conducted themselves at this inquiry.

Manufacturers certainly approached the inquiry in a rather curious manner. While jockeying to acquire a competitive edge in the small domestic market, the New Zealand companies cooperated in their case to the inquiry. Having set up the Central Committee of Tobacco Manufacturers, designed to present a united front against the anti-smoking movement, the companies virtually closed ranks. They presented an air of injured disbelief that 'their' growers could turn on them. Although they had appointed a legal representative, the manufacturers declined to make any submissions until they had heard all other evidence. It was also stipulated that the companies did not wish to be cross-examined on their submissions. This did not prevent Lewis from cross-examining almost all other witnesses. The manufacturers undertook to provide written answers to pre-submitted questions on their submissions and to make a representative available for cross-examination on these answers only. This veto on cross-examination led to a dispute. O'Flynn attempted to find a way around the stipulation and companies protested. Legal opinion was sought. Although the record of the inquiry did not actually limit cross-examination in the way manufacturers desired, an agreement reached 'in chambers' had shown that intention. The restriction on direct questioning led to an exchange of aggressive written questions between the federation and the manufacturers. Each attempted to use the opportunity to reinforce the crucial points of their case.

Both the companies and the federation were critical of the absence of the Tobacco Board, as an entity, from the inquiry. Most board members were present at times in their capacities as growers or company personnel, but the board chairman, Henry Wise, was conspicuous by his absence. To growers this was another illustration of Wise's general unavailability within the industry. Manufacturers, on the other hand, were regretful that Wise's 'enormous background knowledge of our industry and very wide experience in its affairs.'[13] was not being called on. Much of the growth of the industry since the inception of the board could, contended the manufacturers, be attributed to Wise's skill and application. This was a contention which the federation would have hotly debated.

In presenting its initial submission the federation took the opportunity to put forward ideas which had been frustrated at board level for years. The federation claimed that the industry, as presently structured, was 'entirely in the hands of the manufacturers'.[14] Simply by exercising commercial freedom, companies could manipulate production through increasing or reducing the number of contracts offered. In addition, companies retained leverage through the volume of domestic usage. At the time, voluntary usage was over 20 per cent higher than the mandatory requirement.[15] Companies could therefore slash local production by 40 per cent just by reverting to the mandatory 30 per cent usage. The federation again suggested that control of production should be vested in some other authority. It favoured a marketing board or licencing committee which would be independent of companies but would have a strong representation of growers.

Dick Stevens, speaking as federation chairman, now had more than thirty years' experience in the industry behind him. As he outlined the history of the local industry, Stevens was particularly bitter at the new tactics within manufacturers' ranks. For the first time ever manufacturers had appeared before the Price Tribunal, in 1966-67, to oppose the federation's price application.[16] Stevens claimed that Rothmans' case against a price increase had been based on the statement that the season would be 'a bonanza'. The company submitted that yield would be over 1800 pounds per acre, so that growers would automatically be far better off. These statements were made after the tobacco was harvested and the true picture was already known by companies. The season had, in fact, been disastrous for many. Laurence's figures (from the questionnaires and from audited farm accounts) showed alarming reductions in growers' incomes for this season. Some were down by over 30 per cent: hardly a bonanza, whatever the reasons.

Stevens went on to further attacks on companies. He alleged that manufacturer representatives had admitted breaching the stabilisation agreement by allocating contracts above the agreed poundage limit – '[t]hey admitted they had been bad boys'. On the question of standard grades, Stevens considered that company standards fluctuated from year to year. When he had asked at a board meeting what type of leaf manufacturers wanted, 'they said they did not know'.[17] It was little wonder that only a tiny percentage of growers could identify grades and prices being operated by companies at any one time. One long-time grower described this as being 'like a blind man in a nudist camp'.[18]

In response to the periodic claims that growers were inefficient, the federation was adamant that the charge was largely false. If there were inefficient growers this was directly attributable to manufacturers who controlled the industry. 'They chose the growers, chose the land… issued the contracts… financed the growers and… controlled the advisory service on the production of the crop.'[19] If inefficient growers were a problem, the federation asked why manufacturers (Wills) had reduced all contracts by 20 per cent instead of identifying these inefficient growers and ceasing to deal with them. With no pretence at subtlety, Stevens bluntly stated, 'We can no longer continue to play a game whereby the other side makes most of the rules and appoints the referee.'[20] The gauntlet was thrown down for the manufacturers. The inquiry was to be far from the friendly discussion desired by George Laurence.

Noel Lewis proved a formidable advocate for tobacco companies. He made slashing attacks on information and figures produced by all pro-grower witnesses. Lewis claimed the federation's submissions were an open attack on the companies, who had been completely surprised by them. He warned growers it was unwise 'to bite the hand that feeds them',[21] a statement he later withdrew after pressure from the federation and George Laurence. Coming to an area in which he was very familiar, Lewis criticised the price application process, slating severely the method of assessing the value of growers' labour. Attacking the practice of allowing growers a blanket reward for time worked in producing the crop, Lewis claimed it was 'completely illogical and improper to take notice of the producer's own assessment of his time … because it is based on the subjective judgment [sic] of the people involved'. Similarly, Lewis disapproved of attempts to differentiate between the value of family and seasonal labour. 'If family labour is that much more efficient,' he announced, 'then the benefit will show out in the end result.'[22]

Lewis then moved on to the question of the quality of New Zealand tobacco. He declared scathingly that the local tobacco industry was 'probably the worst example of the "she'll be right" attitude'. Growers needed to aim at greater leaf quality and better management practices, Lewis contended, if they were to compete with imported leaf of similar types. Accusing growers of making little attempt to raise their technical skills and knowledge, he referred to a Rhodesian manual on tobacco production, a comprehensive publication of over 400 pages. According to Noel Lewis, this was available to growers in the free section of the Motueka Public Library where, however, 'it rests in peace'.[23]

Clearly, manufacturers were determined to take a hard line on the matter of quality and overproduction. For their part, local growers felt they had been encouraged to produce average, filler-type leaf in ever greater quantities. The superintendent of the Tobacco Research Station, R.W. (Rob) James, believed New Zealand growers were capable of producing an aromatic, high-quality leaf. But James also stated that if quality local leaf could not be produced at comparable prices, it would be more economic to import the total company requirement. Growers were understandably shocked by such pronouncements, which suggested the entire domestic growing industry could be dismantled. The manufacturers were quick to seize on James's comments and to press the federation for a response. The federation reluctantly agreed with the theory but pointed out that overseas exchange was necessary to pay for such imports and that domestic employment should also be considered as part of the economic equation.

Growers took more comfort from James's other contentions – New Zealand producers needed specific standards to work to, and good-quality aromatic tobacco had been produced in New Zealand before the introduction of Virginia Gold. The research station had, in fact, produced such tobacco during the past season. If this type of tobacco could be produced consistently and competitively, companies could look at changing their import policies, making low-quality imports the variable factor in their blends. James also believed New Zealand growers were disadvantaged by the use of inexperienced seasonal labour. This required greater supervision and could affect the eventual quality of the product, a point long made by grower representatives. In addition, James felt the continued cultivation of Virginia Gold, long after the variety had lost popularity with the rest of the tobacco-growing world, was a contributing factor to the production of lesser-quality New Zealand leaf.

For its part, the federation called on local accountants to paint a picture of the parlous state of some growers. At times, these witnesses provided mixed messages. Accountant L.A. (Les) Milnes while standing by the figures produced on the financial position of tobacco growers, also made the comment that 'no more than half ... would be greatly interested ... in regarding their farming as a business operation'.[24] Milnes also thought that the majority of tobacco farmers were not interested in comparing their individual farming operations either with farming in general or with other growers. Despite this, Milnes believed the season just past had been a particularly poor one for many growers. If contracts were cut again, some would 'be in trouble'.

A Nelson accountant, H.G. West, agreed, saying that if the 1965-66 season was going to be the norm, some growers 'would have to give up tobacco growing whether they liked it or not'.[25] West felt there was a point of 'natural justice' involved. Farmers had been encouraged to grow tobacco and had invested in land for the purpose. Figures showed that 43 per cent of current growers had entered the industry within the past six years, the years of intensive company competition for growers. If they were now left with a unit from which there was no return, he felt they had 'a claim for some consideration'.[26]

Noel Lewis disputed this concept. He contended that there were numbers of growers who 'entered the industry with little capital, who have learned and applied the proper methods, who have been efficient and worked hard ... [and were now] ... earning large incomes which other branches of farming cannot produce from much greater areas of land'.[27] Lewis would have no truck with special considerations for those who had not followed this example. The industry, he claimed, would be better off without those who had 'not the slightest idea of the economics of their businesses'.[28] Attacking the figures produced by accountants to back up the federation's case, Lewis described these as unreliable. As they were produced for tax purposes, they were not necessarily a full statement of income. This point, which carried many implications, was somewhat discredited later in the inquiry by the companies' own witness, a farm management consultant. Having slated the method of collecting data from tobacco growers, the consultant then based his comparative case on figures provided by dairy farmers themselves, from their tax accounts.

George Laurence again reprimanded Lewis, this time for questioning the integrity of accountants without being willing to produce the companies' own information. Undeterred, Lewis continued his attack on behalf of the tobacco companies. He now claimed that New Zealand growers enjoyed producer protection unknown anywhere else, through the minimum average price scheme and the minimum usage requirement. The provision by companies of seasonal finance and crop advisers was a further advantage for domestic growers. Lewis returned several times to the matter of accountants' integrity, claiming he had nothing but good relationships with local accountants. He questioned, apparently without irony, only the principle of accepting unverified information from those who had a vested interest in the outcome of the inquiry. Lewis did manage to inject some humour into the proceedings with the statement that there was 'one public accountant who has not been called. I would like her called so that I can express my faith in her integrity'.[29]

The evidence of a local real estate agent, Colin Baas, provided yet another viewpoint on the general position of the industry. With twenty-six tobacco farms on his books, Baas had sold only one in the last three years. Pip-fruit orchard sales had been steady, but the adverse publicity surrounding the tobacco industry in the past two years had definitely affected buyers' willingness to consider purchasing a tobacco unit. Most would like to have the security of a longer contract than the annual arrangement in place at the time. Lenders also liked to know that the terms of the mortgage would be honoured. Noel Lewis worked hard to get Baas to say that the adverse publicity limiting sales of tobacco farms could be laid at the feet of the federation chairman. Baas appeared reluctant to say so outright but his answers to Lewis's questions certainly pointed in that direction. Establishing that Baas was aware of the adverse publicity, which Lewis claimed was emanating from the federation chairman, Lewis then asked: 'You are trying to sell tobacco farms and you are up against the effects of the publicity?' Baas: 'Yes.'[30] Federation counsel made no response to this.

For its part, the federation expressed suspicion regarding the actions of the manufacturers since the beginning of the inquiry. Dick Stevens noted that growers who had previously been refused contracts had since been offered a contract. One such grower had the balance of his outstanding debts paid off by the company as well.[31] O'Flynn alleged that at least part of the extra 50,000 pounds of contract allocated to growers in 1966-67 had gone 'to those growers who were either complaining the most or were in the poorest position', in an attempt to influence these growers in companies' favour.[32] O'Flynn conceded that his remarks regarding company control of the industry had offended manufacturers, but claimed that he had not intended to imply anything sinister in the concept of 'company control'. He felt, however, that 'in a pinch/emergency/crisis there is a natural temptation for the person or party with the control to use it in their own interests first'.[33]

Most of the inquiry hearings were held in the Motueka courthouse and were open to the public. Growers' participation varied. Some attended regularly and took great interest. Others found the whole exercise a waste of time and were sceptical that there would be any great result from what they believed was essentially a political exercise. The continual extension of the inquiry also undermined growers' belief that anything would come of it. As they had been assured of a report within three months, there was understandable derision as the inquiry stretched first into one year and then into another. By 1968 virtually all idea that the outcome would be of any practical use had gone. Those remaining in the industry, now numbering 526, the lowest since 1959-60, were back to the production levels of the late 1950s, both in yield per acre and in total production.

Much of the federation's energy, even during the inquiry proceedings, was spent on battling the Price Tribunal for increased returns for growers, many of whom were now said to be in dire straits. The federation maintained that growers were 'the victim of a war between manufacturing companies'.[34] While companies had increased their profits over the past three years, growers had experienced falling returns. Applying for a massive two shillings per pound increase in 1966-67, the federation claimed this was simply what was justified on the grounds of lower yields, changes in company policies and the need for a higher personal reward for growers. The tribunal responded with an offer of sevenpence-halfpenny, the highest single increase ever. Companies responded by tightening grading standards to give themselves more leeway in accepting leaf. Stevens now claimed that growers were 'running harder to stand still'.[35] After one lengthy Price Tribunal deliberation had produced a meagre 3.8 cent price rise, Dick Stevens commented contemptuously 'the mountain has laboured and brought forth a mouse'.[36]

Family labour remained a critical factor in domestic production. It was claimed that twenty-one of twenty-three wives on farms surveyed in the Dovedale-Woodstock district were working in tobacco for seven consecutive months of the year. The remaining two were under doctors' orders not to work. So too were some of those who were, nevertheless, spending the bulk of their year planting, weeding, hoeing, lateralling, harvesting and grading.[37] This was in addition to their usual daily round of work as rural wives and mothers. In his submissions to the inquiry, Rowly Talbot confirmed that many tobacco wives were also required to provide 'unpaid welfare work' for their workers, in the form of help and advice with personal problems. Farm women, who often worked without wages, were worth at least as much as the casual women workers he employed on his farm, who earned more than £300 in an average season.[38]

The Price Tribunal remained unmoved by such graphic and emotional appeals. It continued to grant less than the application, always maintaining that the federation did not produce sufficient material to back its case. Growers' accounts were submitted annually to illustrate the basis of the federation's price application, but the tribunal restated regularly that the anonymity of such accounts made it impossible to compare one year with another. The veiled suggestion was that the federation was choosing only the accounts of growers which supported their picture of the urgent need for higher prices. One local accountant obviously agreed. A confidential letter to at least one grower from the federation stated that, in the accountant's opinion, 'some of his clients are doing so well that they fear a cut in price instead of an increase'.[39] The federation urged the grower to submit his accounts even if this was the case. When merged with other accounts the effect would be minimal. Another grower was in danger of undermining the federation's case as, despite losing his entire crop in one season, he had 'bought a bigger and better boat!'[40]

Despite the apparent success of some growers, prices rose steadily during the 1960s, although not in line with the federation's applications. Price increases granted for the years 1966 to 1969 totalled approximately eleven cents per pound, a far cry from the twenty cents (2s) claimed in 1966 alone. Even this increase included a five cents per pound compensatory payment granted in 1968 and retrospective to the 1967-68 crop. For two successive years the federation's price applications were delayed by the Price Tribunal's determination to wait for the information on costs it expected to receive from the inquiry questionnaires. Even companies were complaining over the price delay. Wills' manager, Peter Wild, was critical that the price for 1968 had not been announced until five weeks after the crop had been purchased.[41] This left companies to purchase 'blind' without knowing the extent of any supplementary payment they would need to make. Not all growers were convinced of the wisdom of ever higher price applications. A. (Bert) Skillicorn told growers in 1967 that if the price rose by the expected sixpence per pound he believed the companies would 'wipe growers completely'.[42] Skillicorn was expressing the largely unspoken belief that companies had absolute power over the growing industry and that growers should tread warily. The tradition of silence on this point was being broken more and more. J.P.B. (Jim) Hamilton claimed that growers were frightened to speak out because of their financial dependence on manufacturers. Maurice Cederman was again blunt in his approach. Growers were scared, he stated, 'and the people we are dealing with know it'.[43] There was plenty of fuel for the fiery exchanges of the inquiry.

Between the protracted inquiry and annual battles with the Price Tribunal, growers' frustrations were ongoing. Outbursts were often directed at the Tobacco Board and its long-serving secretary-cum-chairman, Henry Wise. In 1967 W. Fry proposed to the federation AGM that it was 'about time Mr. H.L. Wise was sacked'.[44] Similar comments had been made at ward meetings. Wise had attended only one federation AGM since becoming board chairman in 1958 - growers could hardly feel that the nominal head of the industry was personally interested in their affairs. Two new board members, John Hurley and Gerald Hunt, believed the board was unable to reach necessary decisions because the chairman did not cast his vote on vital issues. The resulting deadlock meant the industry was 'left in the air'.[45]

The Minister of Industries and Commerce, Jack Marshall, did attend the 1967 meeting and was forcefully reminded of his duty in the eyes of growers. Dick Stevens, introducing Marshall, had previously described the Minister as the father of the industry, 'and here is your family', he told the Minister, indicating the 150 growers present. Jack Marshall seemed interested in growers' concerns and was apparently supportive of the industry. However, he reminded growers that tobacco growing was a protected industry but that 'protection was not a substitute for efficiency'.[46] At the same meeting, the local MP, Bill Rowling, left no doubt of his position. He proclaimed that growers were not 'lap dogs or puppets' and that the federation's stand at the inquiry was necessary to protect growers' interests. Growers and manufacturers were interdependent and shared equal responsibility to stabilise the industry. Also attending this meeting, George Laurence was given a favourable hearing on matters concerning the inquiry but was unable to say when it would be over.

Discussions with manufacturers on plans for stabilisation were, in fact, making no progress at all. Some growers were without contracts for the coming season. Companies did agree, in August 1967, to allocate contracts on a basis which restored the industry to 10 per cent below the 1966 level. Buying conditions and plant spacing requirements remained the same, and growers were reminded strongly that overplanting would be strictly dealt with.[47] The position remained grave for a number of growers. On many farms, necessary replacement of buildings and machinery was deferred. Even maintenance was out of the question for some.

Early in 1968 the federation presented its own plans for stabilisation to a general meeting of growers. Among the main points was the abolition of contracts, which Stevens described as 'a ball and chain which shackled growers to manufacturers'.[48] The federation proposed giving the Tobacco Board powers to contract with manufacturers for specific amounts of leaf. This would then be allocated to growers on a production quota basis. Standard grades were also proposed so that growers would be aware as they grew what was required, rather than waiting until their leaf was either accepted or rejected at buying time. No agreement was reached with manufacturers on this plan. The only concessions obtained were a guaranteed acreage of 5009, with yield per acre set at 1650 pounds, later raised to 1740 pounds per acre. Company promises to consider federation suggestions on changes to buying procedure in the future were greeted with predictable scepticism by growers. It seemed impossible to break the shackles of the system of annual contracts which were, in effect, the unstable element in the domestic industry.

In each decade of its existence the federation was dominated by one personality. In the 1960s and into the 1970s the dominant figure was Richard Wheatley Staples (Dick) Stevens, a man of intense application and dedication to the industry he chose to represent. Large in stature and stubborn by nature, Stevens was an exacting leader with a mind for detail. He took copious notes of every meeting he attended, often correcting the official record from his own notes or reminding speakers of what they had said. A well-researched speaker, Stevens was easily able to relate to the government officials and industry leaders he was required to deal with in his role as official guardian of tobacco growers' interests. From his own rather tough country upbringing Stevens was, however, scathing of the parasitical nature of city development at the expense of the rural sector. Often angry and frustrated with the power wielded by bureaucrats, he was never outwardly ruffled in dealing with them. Having spent time as a manufacturers' representative, Stevens also understood the dealings of companies and was under no illusions regarding the competitive nature of the industry. Even so, once elected as chairman of the federation he was unswerving in his belief that manufacturers must be forced to give growers a better deal.

On a personal level Stevens was sympathetic to the ordinary grower. In reality, however, he in fact spent so much time on federation affairs that he largely relied on his four sons to carry on the substantial family farm. Even weekends were taken up with discussions with other growers on the back porch of his Riwaka home. Stevens enjoyed the travel and social life that came with the many trips to the capital for board meetings and other federation business. He made many contacts which were to stand him in good stead in his role. Stevens was a respected protagonist in the price wars and company negotiations which were an ongoing feature of the industry. 'Tobacco consumed him' is a common comment now made about him and it is delivered almost as an accolade.[49] With his two terms on the Tobacco Board combined, Dick Stevens was the longest-serving grower representative on that body. He was chairman of the federation from 1966 to 1976, being awarded an OBE in 1974 for his services to the tobacco industry. Even after his reluctant retirement as chairman, Dick Stevens remained a dominant figure until his sudden death in 1979 at the age of seventy-four.

Stevens's reign as chairman was aided by the federation's again deciding, in October 1966, to appoint a full-time secretary-accountant. This time the executive chose British-trained Willis Bond. A mild, courteous man, Bond was conscientious and deferential. He carried out his duties well, being immediately thrown into the midst of preparing material for use at the inquiry. Besides this and the normal secretarial duties, Bond spent much time working on the federation's annual price applications. Dick Stevens had insisted that an accountant was needed to prepare the case for the yearly battle with the Price Tribunal, but Bond's position as an employee of the federation was perhaps detrimental to its case. As early as 1965, T.D. Canton had suggested that the federation should use a Queen's Counsel to present its submissions, believing it was too much to expect the federation to do this. As the Price Tribunal had members of thirty years' experience it was no wonder the federation's case was often 'torn up and thrown in the wastepaper basket'.[50] Canton's suggestion was to come to fruition in the years following the inquiry. Bond remained secretary of the federation until 1972. With the inquiry over, other matters fully occupied his time, especially the production of the *Tobacco Growers' Journal* and the annual allocation of Fijian workers in the early 1970s.

By mid-1968 Laurence was almost ready to submit his report to the Minister. At the federation's AGM, in September 1968, he was able to give the 200 growers present a précis of his findings and recommendations. These largely vindicated the federation's position in addressing the structure of the industry and the perceived control of the companies. Stating that growers had become increasingly subservient to companies, Laurence felt this tendency should be reversed. The contract system, which contributed to this relationship, should be replaced by quotas allocated by the Tobacco Board. The board would then deal directly with manufacturers. Laurence also contended that licences to grow should not be limited to one year but should be continuous until surrendered. Licences should also be transferable, a suggestion which had been wholly rejected by the board in 1966, possibly through manufacturer opposition. Longer-term licences, which could be sold with a farm or transferred to other land, would give growers security and would add value to their farms.

Standard grades were recommended, along with government-appointed classifiers and extension officers to give technical advice to growers. This effectively removed these functions from company control. More research into alternative varieties, along with manuals of growers' instruction, were suggested. The Tobacco Board should be expanded to include management and technical advisers. Laurence's recommendations were 'received with acclamation by the crowded meeting… [and]… he was thanked for the care he had taken with the inquiry'.[51] Laurence's recommendations appeared to address all the major concerns of growers. It would be six more years before any legislation came into effect.

Local MP Bill Rowling was to be a staunch supporter of the industry throughout his long parliamentary career. He warned growers not to 'get carried away if another company should enter the market and try to woo [them] with specious promises'.[52] This was a clear reference to the recent appearance of the British company, Phillip Morris, on the New Zealand scene, having taken over Godfrey Phillips. Rowling saw a danger of the company competition of the early 1960s being repeated to the further disadvantage of growers. From the chair Stevens remarked, without obvious irony, that 'as long as the Central Committee of Manufacturers controlled the industry there would be no competition for the growers' services'.[53] There was again criticism of Henry Wise for not attending the meeting. This time Wise's apology was not accepted – growers' patience was at last exhausted.

Wise, in fact, resigned in May 1969, having fired a parting shot in his final board report. Recognising the crisis of the mid-1960s, Wise nevertheless reminded growers of their advantages under the New Zealand system. In times of crisis, Wise continued, 'blame can be wrongly attributed' and relying on popular means of solving problems might not prove advisable. Giving his report a rather dramatic flourish, Wise concluded, 'A heavy responsibility rests on those within whose hands lies the destiny of the industry.'[54] There is no record of appreciation being expressed for the long tenure of Henry Wise. After thirty-two years' association with the Tobacco Board, his departure was unlamented by growers who had long called for his removal. Wise's suggestion that he could remain as board secretary was not taken up.

Henry Wise was replaced as board chairman by P.B. Marshall, who was said to have a reputation for 'efficiency and courage'.[55] Marshall immediately set about establishing a relationship with the federation, attending the AGM in September. He told 170 growers present that the findings of the Committee of Inquiry would be discussed at board level in order to work out the future of the industry. Marshall believed that the contract system might well disappear, that independent leaf classifiers were needed and that the composition of the board needed to change. Patience, hard work and harmony were needed to reach a satisfactory conclusion.

Patience and harmony were in short supply within the industry. In June 1970 Marshall resigned 'for personal reasons' – he was replaced by J.F. (Frank) Cummings. Price applications remained the main bone of contention. The federation regularly employed Frank O'Flynn to conduct its case before the Price Tribunal and manufacturers now consistently appeared as well. In 1969 the federation succeeded in its appeal against the 1967-68 price. A retrospective compensatory payment of five cents per pound was granted and a further two cents per pound for the 1968-69 crop. But there would be no increase for the 1969-70 season. The federation was reminded that the price for the previous two seasons included the five cents compensation element, and that increases of this size would not necessarily be repeated.

With the arrival of the new confrontational way of dealing with the price applications, the old-style paternal relationship between companies and growers was also shifting focus. Long-time company men such as Charles Pethybridge and Phil

Littlejohn retired, and more hard-headed business tactics became common. Local company personnel developed close and mainly friendly relations with their growers, but their position was often that of 'piggy in the middle'. Many, like Peter Wild and Bert Black (Wills), Eddie Bradley (Godfrey Phillips) and Bill Boyden (Rothmans), were long-standing members of the local community. Under the new, somewhat adversarial tactics, they could find themselves viewed with some suspicion when required to deliver company policies on leaf quality and buying procedures.

Growers themselves, while hoping for a favourable outcome from the inquiry, were still required to keep their minds on the job of producing 'the weed'. Modernisation had proceeded apace through the early 1960s. The majority of farms were now using at least some of the innovations of the past decade. Irrigation units were commonplace as were down-draught kilns, tying machines (now separate from harvesters) and mechanical harvesters. With the spread of mechanisation very little of the tobacco culture process, apart from grading, remained as highly labour-intensive as before. Companies also changed some of their methods during this time. Rothmans trialled the buying of loose or unhanked tobacco. This was said to save growers two cents per pound in labour costs – growers, however, received two cents per pound less for tobacco sold loose.

Despite these innovations, large numbers of seasonal workers were still required for the annual harvest and shortages of workers remained a seasonal headache. There were headaches, too, with some of the workers who did come to the district. Weekend parties created problems for the placement officer on Monday mornings as the effects of sackings and defections began to appear. Many growers had prohibited drinking in their baches but a small number were said to be undermining the position by neglecting to place any controls on the behaviour of their workers. With the ongoing problems of unruly workers thought was beginning to turn towards using imported labour, possibly from Fiji, to provide a stable and more controllable workforce.

Curiously, while the federation consistently fought growers' battles on all fronts, for those on the tobacco farm the picture was not always as serious or unhappy. Many enjoyed the annual influx of workers as a glimpse of other places, often other countries. For those confined to their own farms for much of the year the harvest season provided an opportunity to talk with, and enjoy the company of, a variety of people from many walks of life. Happy times were spent over the teacups at smoko and at end-of-harvest celebrations. These are recorded in numerous photographs and in many farmers' memories. Some formed lasting friendships with visiting workers. Others enjoyed the often unconventional antics of seasonals, both their own and others.

Harvesting itself was often a congenial occupation as it took place during the usually very pleasant summer weather of the area. Along with workers, many growers enjoyed the general atmosphere of activity and productivity which the season generated and a great deal of fun was had by all. As well, many growers were unaware of, or uninterested in, the political in-fighting within their own industry. Women, especially, were given little idea of the nature of decisions which were being made and which could have significant impact on their livelihood. Even the feminist movement of the 1970s did not bring any women into the federation fold – the vast majority never attended an annual meeting. For these women, tobacco harvesting brought not only a heavy workload but a social experience, often enjoyable, sometimes not. Many now look back on this with warmth and an affection for the uniqueness of their occupation.

Tobacco women were not wholly unrecognised for their part in the production of the crop, but it was difficult for them to be adequately rewarded financially. In a letter to the *Tobacco Growers' Journal* in May 1967, a grower graphically described his wife's worth:

> She works in the seedbeds, and cultivating the crop; she helps with harvesting and grading; takes her turn with the kilns; keeps my accounts, makes tea for the workers, advises and helps the female seasonal workers, and does countless other jobs. She knows what has to be done and can be trusted to get on with it swiftly and without supervision. She is worth at least £500 [per year] to me and it seems unfair that I can only pay her £156 without losing the special tax exemption for her.

The grower was advised of ways of adjusting this anomaly within the tax system but, like many other rural women, tobacco wives could never be fully recompensed for the contribution they made to the viability of their farms.

While little changed for women the research station did see changes in personnel during the 1960s. Bert Thomson retired in 1966, after twenty-five years' work in the local industry. He was replaced by Rob James, who had been an assistant chief research officer in the Zambian tobacco industry and who had extensive experience in tobacco growing in that country. James walked right into the middle of the controversies of the Committee of Inquiry. Almost immediately, he found himself sandwiched between growers and manufacturers. From this beginning some growers were a long time in accepting some of his ideas on tobacco cultivation, varieties and the general structure of the industry. Today some readily concede that Rob James 'had it right'[56] in many respects.

Also in the 1960s came Ethena A. Walker, specialising in plant diseases. She had spent seven years working on tobacco diseases at the Cawthron Institute in Nelson before coming to the research station in 1962.[57] Verticillium wilt was still a major concern, although research into the disease was intensive. Field trials with the chemical chloropicrin were conducted in 1964. This involved injecting the chemical into the soil and was estimated to cost the farmer £100 per acre. The disease remained unconquered, however, and was the focus of research work until the late 1970s. Following the retirement of Ethena Walker in 1968, the appointment of Dr A.S. (Anoop) Bedi, a plant breeder trained in Britain, reinforced the shift in the focus of the Tobacco Research Station away from research into disease towards developing specific varieties for the area. Bedi's replacement, R. (Ron) Beatson, continued this trend. Beatson was born into a tobacco-growing family and had completed his doctorate in tobacco plant breeding. He is widely credited with the development of new, successful varieties in the 1970s and 1980s.

As always, no amount of mechanisation or research could combat the ever-present seasonal hazards. In 1965 hail damage in early March destroyed over 60,000 pounds of tobacco. This was followed by severe frosts less than two weeks later, which caused the greatest single loss of tobacco leaf in any one season – 656,000 pounds, worth over £145,000. Two years later the crop was similarly affected. Heavy flooding in December 1966 was followed by widespread hailstorms in January 1967 – more than 400 acres of tobacco were damaged or destroyed. 'All we need now is a good frost to finish things off,' commented one farmer.[58] The frost came early in February, devastating crops in Orinoco, Stanley Brook and Tapawera.

The industry had now lost most of the marginal growers and all of the community organisation contracts. As the tighter standards of the companies bit deeper, more and more farmers for whom tobacco had been a valuable supplement to their annual income opted out of the industry. The areas most affected by the shrinkage in acreage were mainly the outlying areas, often classed as marginal. Wai-iti was the most dramatically affected area, losing 57 per cent of its 1963-64 acreage by 1970. The Moutere Valley, Waimea and Dovedale all lost more than 20 per cent in the same period. By contrast, acreage in the Motueka Valley and Riwaka decreased by less than 10 per cent. In total, area in production shrank by 14 per cent in the second half of the decade. Surprisingly, despite disappointing prices, seasonal disasters, labour problems and the shrinkage of the past five years, the growing industry was hanging on in relatively good heart as the decade ended. Production was rising again and Laurence's report was still expected to be the basis of favourable legislation.

The findings and recommendations of the Committee of Inquiry were finally released in September 1969. Laurence recommended more defined duties and powers for the Tobacco Board in order to avoid the 'unnecessary disturbances' within the industry, which he believed had been caused by the 'policies and practices of tobacco manufacturers'.[59] Licences to grow tobacco should be continuous and transferable and contracts, or production quotas, should be allocated by the board. Laurence also recommended that the board should closely monitor tobacco stock levels to avoid the setbacks caused by overproduction which he laid squarely at the feet of manufacturers. The board should actively promote and encourage domestic production, and should provide specific manuals on soil conditions, farm practices and management. Field advice should be offered by Department of Agriculture officers, to give farmers free choice in obtaining advice. Leaf classification should be carried out by government-appointed classifiers.

Laurence suggested that the price of domestic leaf should be set by the Minister of Industries and Commerce, after receiving a report from the Tobacco Board. This brought criticism from Bill Rowling, who saw danger in the possibility of political pressure affecting the process. Despite his reservations on this point, Rowling was to battle constantly over the following five years to introduce legislation which would embody the intent of the inquiry. The 1960s came to a close with the industry in a curious mood of optimistic resignation. The decade had been typical of the see-saw life of the tobacco-growing community – some were now out of the playground for good. Those remaining would come to see the 1970s as crucial in determining the future of their industry.

NOTES

1. Figures prepared by Tobacco Growers' Federation secretary Bond, 31.1.67, to support 1967 price application.
2. Tobacco Board report, 1967.
3. Tobacco Growers' Federation Annual Report, 1967, p.3.
4. *Nelson Evening Mail*, 26.8.66.
5. Ibid.
6. Transcript of the Public Hearings of the Committee of Inquiry into the Tobacco Growing Industry 1966-68, p.1.
7. Ibid, pp.5-6.
8. Ibid, p.11.
9. From Stevens's briefing notes to O'Flynn, p.5, Tobacco Growers' Federation files.
10. Transcript of the Public Hearings of the Committee of Inquiry into the Tobacco Growing Industry 1966-68, p.120.
11. Ibid, p.121.
12. Ibid, pp.185-86.
13. Manufacturers' answers to Tobacco Growers' Federation questions, 27.11.67, federation files.
14. Transcript of the Public Hearings of the Committee of Inquiry into the Tobacco Growing Industry 1966-68, p.19,
15. In addition to the mandatory 30 per cent usage, companies' voluntary domestic usage had brought the total domestic usage to over 50 per cent. See board reports and inquiry transcript.
16. Transcript of the Public Hearings of the Committee of Inquiry into the Tobacco Growing Industry 1966-68, p.24. While companies had previously accepted the decisions of the Price Tribunal, there was widespread suspicion of behind-the-scenes influence by manufacturers.
17. Transcript of the Public Hearings of the Committee of Inquiry into the Tobacco Growing Industry 1966-68, pp 29-30.
18. *Nelson Evening Mail*, 21.7.66.
19. Ibid. This led to the claim that the membership of the Tobacco Growers' Federation was, in reality, 'chosen by the manufacturers'.
20. *Nelson Evening Mail*, 3.3.67. Stevens was initially reported as having used the words 'breaks the rules', eliciting severe criticism from manufacturers. The inquiry transcript shows 'makes the rules' to be his actual statement.
21. Ibid, 29.8.67.
22. Transcript of the Public Hearings of the Committee of Inquiry into the Tobacco Growing Industry 1966-68, pp.192-93.
23. Ibid, p.194.
24. Ibid, p.108.
25. Ibid, p.95.
26. Ibid, p.97.
27. Ibid, p.201.
28. Ibid.
29. Ibid, p.156.
30. Ibid, p.184.
31. Handwritten notes from Stevens's personal papers, Tobacco Growers' Federation files.
32. Transcript of the Public Hearings of the Committee of Inquiry into the Tobacco Growing Industry 1966-68, pp.185-86.
33. O'Flynn's handwritten notes in Stevens's personal papers, Tobacco Growers' Federation files.
34. Tobacco Growers' Federation representations to Price Tribunal, 19.3.69, at Motueka. Federation files.
35. *Nelson Evening Mail*, 20.7.67.
36. Ibid, 21.9.68.
37. Tobacco Growers' Federation submission to Price Tribunal, 19.3.69, federation files.
38. Transcript of the Public Hearings of the Committee of Inquiry into the Tobacco Growing Industry

1966-68, pp.176, 179.
39. Letter from Tobacco Growers' Federation to unidentified grower(s), 18.3.68, federation files.
40. Memo from Tobacco Growers' Federation secretary to Dick Stevens, 20.6.67, federation files.
41. *Nelson Evening Mail*, 23.9.68. In the 1970s it became standard practice to consider the price application after the buying season had ended, with companies making preliminary payments based on the previous season's price and 'top-up' payments once the price was announced.
42. Ibid, 24.6.67.
43. Ibid, 24.8.66, 16.8.67.
44. Tobacco Growers' Federation minutes, 8.9.67.
45. *Nelson Evening Mail*, 29.11.67.
46. Tobacco Growers' Federation minutes, 8.9.67.
47. Letter from central committee of Manufacturers to Tobacco Board, 9.8.67, Tobacco Growers' Federation files.
48. Tobacco Growers' Federation minutes, 22.3.68.
49. Interview with John Hurley, 1.8.96. Affirmed by many other growers.
50. *Nelson Evening Mail*, 6.7.65.
51. Tobacco Growers' Federation minutes, 12.9.68.
52. Ibid.
53. Ibid.
54. Tobacco Board report, 1968.
55. Tobacco Growers' Federation minutes, 3.6.69.
56. Interview with former grower, federation chairman and Tobacco Board member Gerald Hunt, Motueka, 30.8.96.
57. *New Zealand Tobacco Growers' Journal*, January 1969, p.25.
58. *Nelson Evening Mail*, 7.1.67.
59. *The Economic Position of the Tobacco Growing Industry in New Zealand*, Report of the Committee of Inquiry, Government Printer, 1971, p.132.

CHAPTER SIXTEEN

Bill and the Bill

Like virtually every other decade, the 1970s were a roller-coaster ride for growers. It was also a tough time for manufacturers as they battled changing international trade patterns and faced constant pressure from the growing sector.[1] In addition, successive governments proved crucial during the 1970s. Growers had been warned that legislation enacting the findings of the Committee of Inquiry might take several years. Growers placed their faith first in the National government, which failed to act. They then turned to their local MP, Bill Rowling, who introduced a series of private member's bills built around the conclusions of the inquiry.

Early in his short tenure as board chairman P.B. Marshall met with the federation. Marshall indicated that, while he intended to implement the findings of the inquiry, any recommendations from the board must be unanimous otherwise it was 'like a toothless old dog'.[2] Growers made their position quite clear at this meeting with Marshall. Manufacturers could not be trusted, they claimed, and gave examples to back this up. The federation executive reiterated to Marshall its stand on the various concerns covered by the inquiry and went on to say that if the board did not act quickly the federation could consider taking unilateral action.

Marshall's task at board level was to prove more difficult than he expected. Over the following few months, discussions at board level became more and more acerbic. Manufacturers began to question the expenditure of the federation which was funded through an annual board grant from grower levies. Company representatives stood firm on the questions of independent leaf classifiers and a change in the control of the industry. The manufacturing sector was fundamentally opposed to both. Marshall resigned in frustration after less than a year in office. He described the grower members of the board as 'intransigent' when they also refused to move on these two issues.[3]

Marshall's replacement, Frank Cummings, was a former Comptroller of Customs and was said to be conservative in his approach. K.D. (Ken) Butland, a Rothmans representative on the Tobacco Board, suggested that all restrictions on the industry should be abolished, and companies could contract with growers 'over the fence'.[4] Growers stood firm. Cummings believed growers should compromise but Dick Stevens retorted that this would not be compromise but surrender.[5] Stevens was, in fact, holding private discussions with some company representatives and thought that the 'ice was beginning to crack'.[6] If growers stood firm they would win out. For his part, Cummings was yet to be convinced of any instability in the industry. He should have had no doubt of the strength of growers' feelings when he visited the area in November

1970. Criticism was levelled at almost every company activity. Along with their long-standing complaints against price and buying practices, growers now claimed that company field officers were not as well qualified as companies made out. Cummings was informed that there were 'more members of the Federation with degrees and diplomas in Agriculture and Horticulture than there were in the combined manufacturers' field officers'.[7]

Growers were also suspicious of the companies' opposition to independent classifiers. Manufacturers claimed to be against the idea on the basis of the cost to the industry, but growers believed that there must be another reason for manufacturers preferring to classify leaf themselves. Companies were not benevolent societies, said the federation. Independent classification 'would reveal that they had been getting high-quality New Zealand leaf cheaply'.[8] Companies naturally contested this view. Early in 1970 Butland visited both John Hurley and Dick Stevens, informing them that New Zealand leaf was close to imported leaf in price and stressing that companies would resist any attempts to increase government control of the industry.[9] Butland also hinted that manufacturers might try to split growers in order to weaken the federation's opposition, adding strength to suggestions that companies had been involved in the divisions of the 1950s. The confrontational attitudes evident in the proceedings of the inquiry were flourishing in the early 1970s.

The annual price-fixing round provided another source of friction. Manufacturers appeared, without prior notice, before the tribunal in December 1970, with a lengthy submission for a price increase five cents below the federation's application. When the hearing resumed in the new year, the companies withdrew much of their submission which the federation had revealed as containing incorrect information. At board level discussions of the inquiry report continued. Agreement was reached on all except the vital issues of classifiers, board control of the industry and a levy on manufacturers. Opposing the federation's case for these changes, both Cummings and the company representatives claimed that most growers were content with the present system and had happily signed their contracts.[10] Worse was to come for growers. The Minister of Industries and Commerce, Norman Shelton, claimed that growers were 'only interested in what went in their pockets'.[11] Shelton's allegation could only be explained by the federation as coming from an almost complete lack of knowledge of the industry he was nominally overseeing. Within two months the federation reported a better reception from their Minister and solid support from his department. Growers had also reacted to the suggestion that they were happy with the current set-up of the industry by signing protest letters to the board in large numbers,[12] although it is unlikely that this outpouring was spontaneous.

Much of growers' concern over the 'current set-up' stemmed from its lack of stability. In 1970 the Tobacco Board financed three of its members, Dick Stevens, George Laurence and L.P. (Peter) Hamlin (Rothmans), to visit the tobacco-growing areas of Australia to study the effects of the stabilisation scheme introduced in 1964. Dick Stevens reported to the federation that this scheme was so successful that all parties had requested that it be continued for a further five years. Australian growers enjoyed protective duties three times higher than those in in New Zealand, enabling manufacturers to pay higher prices for Australian leaf. The leaf was sold at auction, with minimum prices and a grade-price schedule, and manufacturers received a rebate on import duty if they used a specific percentage of domestic leaf. Stevens claimed the Australian government 'considered people more important than economics'.[13]

Dick Stevens produced his own observations very shortly after returning from the trip. But the official report of this board delegation was long in coming, as Laurence, in his deliberate and painstaking way, took his time. Hamlin produced his own report, having cut short his part of the trip. His comments were largely a company critique of the official report, which he considered had introduced comparisons with the New Zealand situation rather than simply reporting on the Australian system. He also felt his discussions with company counterparts had given him information not available to Stevens and Laurence. The official report was kept confidential to the Tobacco Board and, three years later, had not yet been discussed.[14] Many of the points it raised were, however, addressed in the drafting of Rowling's private member's bills.

Bill Rowling introduced his first private member's bill in 1971. The opposing attitudes within the industry were reflected in political divisions during debates in the House. The Minister of Industries and Commerce, Shelton, complained bitterly that he had not been informed of the intention to introduce the bill, even though he had been involved in numerous discussions with growers during the past few months. Growers found this hard to swallow. Officers of Shelton's department had attended their 1970 AGM, where the proposal had been openly announced. Shelton appeared to believe that the difficulties of the industry could be resolved at board level and that, as seventeen of the nineteen recommendations of the inquiry had been agreed upon, there was little need for legislation.[15] In response to a question on what the government intended to do if total agreement was not reached, Shelton replied, 'Well, we will make up our minds.'[16]

The government's case was both defensive and somewhat weak. The Minister of Justice, D.J. (Dan) Riddiford, certainly showed a woeful ignorance of both the industry and the provisions of the proposed legislation. Attempting to back up Shelton's case, that the bill reflected only growers' needs, Riddiford claimed that the proposals would give growers domination. He seemed to believe that growers would now outnumber manufacturers on the Tobacco Board. Going further, Riddiford claimed growers had a captive market in New Zealand and there 'was a time when growers had matters too much their own way'.[17] Pressed to say when this was, he admitted he did not know, but mentioned the company competition of the early 1960s. Riddiford concluded by describing producers' interests as 'often narrow' – he had added little to the rational discussion of the issue.

Speaking immediately after Riddiford, the member for Nelson, Stan Whitehead, regretted that government members were not taking the issue more seriously. Referring to the previous speaker as 'pitiful', he said he thought it would have been better if Riddiford had not spoken at all. Whitehead traced the shrinkage of the industry and complimented growers and researchers on increased yields. Growers did not want their industry to be 'at the mercy of the manufacturers'[18] as it had been in the 1960s. This was disputed by the interjection of Shelton who claimed that growers were not at the mercy of companies and that the 1960s were over. The rather unedifying debate wound on, with Labour members obviously far more informed on the subject than their National counterparts. In the end, although the bill was read for a first time, it was not sent to a select committee and received no further consideration during the parliamentary term. While some government speakers had hinted that Shelton was planning his own legislation, there is no evidence of this. The tobacco growers' former advocate, Keith Holyoake, did not speak in the debate.

Dissension continued unabated at ground level. In 1971 companies introduced a change to their contract basis, offering growers contracts based on a quota system. Rothmans led this move, basing its contracts on an allocated poundage. This type of control, along with the tighter buying standards introduced in the 1960s, had the effect of cutting the overall amount of tobacco which growers could expect to sell. At first the grower representatives on the Tobacco Board refused to be involved in this exercise, which appears to have been sanctioned by a board majority.[19] All the old difficulties of overplanting to achieve a greater amount of higher-priced leaf to sell, and of the consequent overproduction, again became a concern.

As well, the quality of New Zealand leaf became a point of contention once more, with Butland claiming that domestic leaf contained unacceptable levels of tar and nicotine.[20] This was described by another manufacturer, M.W.M. (Michael) Rouse of Phillip Morris, as a red herring. Rouse stated that his company's two best-selling cigarettes contained the highest percentage of New Zealand leaf.[21] Undeterred, Butland continued his head-on attack. He warned that he would propose to the board that finance to the federation should be 'reduced or cut off'.[22] Butland's comments and general attitude portrayed the extent of distrust and antagonism felt by a number within the industry and illustrated the difficulty of the board in bringing together these often totally opposing factions.

Amidst the continuing disagreements between the federation and manufacturers, rumours began to circulate that some companies were considering relocating to Australia.[23] Dick Stevens was careful to explain the commercial reasons behind such a move but growers obviously saw it as another of the pressures they were subject to in the atmosphere of tension. George Laurence agreed. He felt that companies would exert a great deal of pressure and subversion 'out on the farm'.[24]

In a practical sense the industry underwent major changes in the 1970s. The average acreage of tobacco per farm increased, mainly in an attempt to reduce costs of production and to overcome the effects of changes to company policies. Mechanisation was gradually reducing the level of labour required to handle the crop – mechanical harvesters, tying machines, travelling irrigators, and changes to kiln technology all required less handling than previous methods. Considerable labour was still needed, however, and growers had become increasingly dissatisfied with New Zealand workers through the 1960s. This continued into the 1970s. One grower declared indignantly that he was sick of the industry having to rely on 'layabouts, bums and tourists'.[25] Male workers in particular 'came and went with exasperating rapidity'.[26]

In an effort to counteract the instability and unreliability of this type of worker, the federation, through Gerald Hunt, initiated a scheme to bring in a number of Fijian workers for the seasonal harvest. After considerable behind-the-scenes work, 142 Fijians were flown to Nelson in January 1971. For most, it was their first taste of tobacco work and the often unexpectedly different New Zealand climate. Community groups found themselves providing heaters and warm clothing for a number of unprepared, cold Fijians. Growers who were allocated Fijian workers were required to advance the cost of airfares, which could then be recovered in weekly deductions from wages. Most were delighted with their imported labour force, finding them far more stable and reliable and unlikely to leave before the end of the harvest season.

The majority of Fijian workers came to the area with the intention of sending their earnings home to Fiji and this was not appreciated by the local retail community. Accustomed to the more free-spending Kiwi workers, retailers felt threatened by the possibility of a workforce which spent little locally. The federation attempted to reassure shopkeepers that Fijian workers simply had a different pattern of spending. The scheme operated for ten years, bringing in up to 300 Fijian workers each season, and included a training element for a small number of Fijians who were employed in their domestic tobacco industry. During these years the local public became accustomed to the presence of the mostly courteous, well-behaved visitors, although it is doubtful that they received total acceptance. Some growers have maintained contact with 'their Fijians' and have visited them while on holiday in Fiji, but none of the hundreds of Fijian workers appear to have returned to settle in the Nelson area. Increasing unemployment within New Zealand placed pressure on the organisers as the 1970s wore on. After 1981 they were no longer able to justify the need for outside labour and the scheme was discontinued.

At the research station Rob James had moved the focus of research more towards plant breeding. James sought always to develop a variety which was resistant to the greatest number of diseases while producing high-quality tobacco under New Zealand conditions. Coming up against companies which, in his opinion, had no idea what quality of leaf domestic growers were capable of producing, James also had to contend with growers who were reluctant to disregard the instructions of their company field officers. From his experience on the African continent, where growers' returns were solely dependent on quality, James found it difficult to contend with the New Zealand situation. He felt that the contract system and the district average price combined to suppress moves towards better quality. James came to view the New Zealand tobacco system not as an industry but as a three-company domination, based solely on the legal requirement to use 30 per cent local leaf in manufacturing operations.[27] New Zealand growers, he contended, did not regard their tobacco as belonging to themselves, but rather as company property, from seed to graded leaf. Growers therefore were, on the whole, not interested in improving quality. As long as the company financed them and directed them in the type of cultivation and leaf production they supposedly required, many growers could see little rhyme or reason in drastic changes to the system.

In the face of significant scepticism James set up district trials, using plots on farms in localities around the whole growing area. In these trials, research station staff planted, cultivated and harvested a number of varieties in order to assess their performance under differing conditions. Farmers generally regarded these trials with curiosity. Not only were the varieties different from what was grown on the farm, but cultural practices also varied. Many growers awaited the results of the research plots with a modicum of derision as they observed procedures which were markedly different from their own. The results often surprised them.[28]

James and his research team tried to maintain an impartial stance. This often meant walking a tightrope between pressure from the federation in support of its contention that growers already produced quality leaf, and indifference from manufacturers who appeared ignorant of the potential of the domestic industry. With many of his statements apparently critical of growers, and with his attempts to introduce 'foreign' ideas, Rob James was a controversial figure within the industry. The situation was often aggravated when James was called in for discussions with the Price Tribunal. Consistently asked to advise the tribunal privately on the quality of local leaf, James came under fire on a regular basis from both sides of the debate. At a general meeting of the federation in December 1977 motions were passed to have James removed or publicly censured. These sentiments were reputed to have stemmed from ward meetings, where growers had expressed anger at James's comments and a lack of confidence in his advice. The federation had no power to have James removed or transferred, but some 'hot-headed persons had taken the bit between their teeth'[29] and the motions to petition for James's removal were pushed through the general meeting.

The executive wisely sought legal opinion on the matter. It was advised that the suggested course of action could involve the federation in legal action and heavy damages. If it could be proven that the statements had been prompted through 'ill-will or spite' the damages could be even heavier.[30] The executive was, in fact, worried that those who had pressed the matter through the general meeting were not necessarily genuine in their intentions and had appeared to take a frivolous approach.[31] James himself was concerned that the proposals left him defenceless as he could not reply to the critics. He considered his career was in jeopardy and would have liked open discussion on the problems.[32] The federation advised the members concerned that it would not be acting on the motions but that the president would have confidential discussions with the Director-General of the DSIR.[33] Whatever the tone of these discussions, the matter lapsed and James retained his position.

The concentration on development of good New Zealand varieties came to fruition in the late 1970s, with the introduction of Kuaka strains based on the narrow-leafed Hicks variety. Developed under the direction of Ron Beatson, these New Zealand-bred varieties were highly resistant to verticillium wilt, black root rot and mosaic and, along with improvements in fertilisers and farm management practices, raised yields to unprecedented levels. The new varieties were especially suitable for the new bulk curing process which became widespread during the late 1970s and 1980s.[34] While never totally accepted by local growers, Rob James had vindicated his belief that, with correct cultural practices and varieties, New Zealand growers could produce quality tobacco leaf reasonably competitively.

With the change of government in 1972, Bill Rowling was finally able to introduce his bill with confidence that it would become law. Much of the drafting was done by Lindsay McKay, of Wellington, under the auspices of Frank O'Flynn who was by now a member of Parliament himself. Again there was party-political division on the matter with National now in opposition. This time Holyoake did speak but dwelt mainly on his previous role in the passage of the 1935 legislation. He hoped only that whatever resulted from the proposed legislation would make 'at least some contribution to ... what has always been, and I fear will always be, a difficult industry to administer'.[35]

Rowling's proposals included most of the recommendations from the inquiry report. Of greatest impact was the proposal to abolish the contract system which would be replaced by a quota allocation. Licences to grow would no longer be tied to annual contracts but would be continuous and transferable. There was provision for the reconstitution of the Tobacco Board. It would have extended functions and powers, including the supervision of the quota system through a Tobacco Quota Committee which would allocate basic individual quotas to growers. The price of domestic leaf would be set by the Department of Trade and Industry, leaf would be classified by independent classifiers, and companies would be required to pay a levy on all leaf used in their manufacturing operations. Manufacturers were also to be required to indicate their needs for the coming five years. The mandatory usage provision was retained and Rowling proposed to extend this from 30 per cent of total usage to 30 per cent in each brand. Finance for growers would be provided through the setting up of a Rural Bank, thereby removing dependence on company finance.

The referral of the bill to a select committee provided yet another opportunity for growers and manufacturers to lock horns over their areas of disagreement. For its part, the federation was delighted with the proposed legislation which offered most of the features they had expected to flow from Laurence's inquiry. Manufacturers, who at first presented a somewhat cursory submission, quickly gathered their forces, resisting and challenging the bill's proposals. They expressed alarm at proposals to abolish the contract system, introduce independent classification and enforce a statutory obligation to purchase specific quantities of domestic leaf. Such measures would affect their 'means of regulating the production and sale of each season's crop'.[36] In response, the companies fought hard to ensure that any new system retained the central elements of the old. Rothmans, for example, while suggesting measures which apparently agreed with the bill, proposed subtle changes which would, in effect, retain the status quo. The companies strongly defended the prevailing contract system. They claimed that it preserved an historic continuity in relationships, enabled the provision of finance and technical advice, and generally preserved 'mutual trust and confidence between manufacturers and growers'.[37]

The federation found this argument difficult to treat seriously. The inquiry had labelled the relationship of grower to company as 'subservient' and the federation asserted that the contract system was probably the greatest cause of this. It lay at the heart of the lack of confidence and trust between companies and their growers. The concept of continuity was disputed on two counts. If it was not in the best interests of growers it could hardly be presented as a virtue of the contract system and, while continuity may have existed prior to 1968, since then the practice of transferring growers between companies had become common. In addition, the federation claimed these transfers were often without the grower's knowledge.

The federation also considered Rothmans' suggestion of a grower board, to negotiate with companies on a 'contractual basis', to be absurd. Such negotiations should be between equal parties, but the federation contended that there was little possibility of equality while companies could refuse to purchase any leaf at all, and while growers' livelihoods depended on growing and selling regardless of conditions and prices. Without a statutory requirement for manufacturers to purchase a specific quantity of the crop, growers would have even less protection than under the current system. The federation maintained that the suggestions put forward by manufacturers could only do harm 'to a relationship which is at present precarious enough'.[38]

The final version of the 1974 Act was much in line with what had been introduced. Although the mandatory overall percentage of 30 per cent was retained, Rowling was unable to introduce the requirement for each brand of cigarette to contain 30 per cent domestic tobacco. Similarly, the concept of a levy on all tobacco used in manufacture, including imports, was discarded. However, manufacturers were required to match the levy on domestic leaf already paid by growers. Manufacturers were also required to indicate their future needs on the basis of definite orders for two years, minimum requirements for the following year and estimated needs for two further years. Growers could now see the probable production figures for five years into the future. With the Tobacco Board restructured, the contract system dismantled, mandatory forward ordering and independent classification of domestic leaf, the federation appeared to have achieved much of what it set out to get.

For growers it had been a long haul from the trauma of the mid-1960s to the passing of the legislation. Now there was hope that the industry would at last be placed on a solid footing with a more equitable sharing of power and a workable means of stabilising production. Company reaction varied. Rothmans was delighted with the measure to remove the financing of the industry from companies and ceased the practice almost immediately.[39] Having offered finance mainly through bank guarantees and securities over chattels, the company moved quickly to transfer, or end, these arrangements. In some cases outstanding debts were written off, but other growers found themselves in difficulties as they awaited confirmation that the Rural Bank would take over their financing. W.D. & W.O. Wills, which had traditionally provided direct finance to its growers, continued to offer this service for a further two seasons.

Almost from the beginning the Act was controversial in its impact. Growers' early hopes of radical change were eroded as manufacturers manoeuvred to maximise the new situation. The election of a National government in 1975 removed Bill Rowling's ability to ensure the true intention of the legislation was carried out. The new structure of the board did not, in the event, reduce tension between growers and manufacturers. Although the mixture had changed, the faces around the board table were largely the same. The younger and newer growers on the board felt the time had come for a change in tactics in dealing with manufacturers. Dick Stevens would have none of this. The tension between the traditionally hostile tactics, and the idea of using the board as a means of exchanging views, was inexorably leading to the end of Stevens's long dominance within the industry. Some instability was also created on the Tobacco Board with the change of government in 1975. The Labour-appointed chairman, Sir Jack

Harris, served only a few months before being replaced by S.F. (Jack) Ashby, who was to be chairman until his sudden death in June 1978.

The transferable quota system, designed to provide growers with some security and to recognise the value of tobacco farms and their particular equipment, turned into a nightmare. For years companies had offered contracts virtually where and as they chose. Now the task of allocating quotas within the manufacturers' district order fell to the three grower members of the board and the board chairman. Hours and days were spent attempting to fairly allocate the required poundage. The committee juggled the domestic order with growers' preferences of which company they wished to grow for, and how much they wished to grow. Mountains of calculations, in the pre-computer era of the 1970s, simply proved to be a poor use of energy and time – at times the committee found itself deliberating for hours over a few kilograms of tobacco.[40] Few were completely satisfied. In some instances the procedure led to appeals and even court cases as growers disputed the quota committee's decisions.

In addition, just when growers had been given some of the measures so long awaited, problems arose in the practical sense. The growing industry failed to produce the annual company requirements for several seasons in a row. The 1976-77 season was disastrous in this respect. Production was over a million kilograms short of the domestic order after cultivation and harvesting were completely disrupted by higher than usual rainfall, fewer sunshine hours and low average temperatures.[41] Companies consequently reduced their annual orders. Growers now felt they were penalised for low production in a bad year but unable to hold over leaf from a good harvest. The federation claimed the manufacturers were forcing the domestic order down in order to import cheap tobacco. Companies disputed this, producing figures to show the relatively stable usage of both imported and local leaf. According to these figures, there had been little dramatic change in the usage of either category since 1961.[42]

The most dramatic change, in fact, was in the market share of the respective companies. In the mid-1950s W.D. & W.O. Wills was said to hold 80 per cent of the New Zealand cigarette market. By 1972 Rothmans had secured 73 per cent of this market, with Wills languishing at 19.5 per cent. Since Rothmans used a lesser percentage of local leaf, its rapid growth in the marketplace affected the overall amount of domestic leaf required by manufacturers. Even this market dominance was a mixed blessing for Rothmans as much of its market share relied on brands that were either declining in popularity or remaining static.[43] While various possibilities were open to manufacturers to address their own marketing problems, growers had only the government to rely on to improve their position. The abolition of the annual contract system had been intended to remove company control of the growing industry. In reality, this had been replaced by control through the annual domestic order, entirely based on company requirements, which appeared to be declining. By 1980 the domestic order had reduced to 2.8 million kilograms, down from contract production of 3.8 million kilograms in 1970.

Grower numbers and acreage also fell steadily throughout the decade. Between 1970 and 1979 grower numbers almost halved, and acreage was down by a quarter. The years of under-production were then compounded by several seasons of production above the domestic order, as the new varieties specifically developed for New Zealand conditions began to take effect. Growers were now in a cleft stick. Finally producing quality tobacco in consistent quantities, they found, under the new arrangements, that companies could not be compelled to purchase any particular amount of their crop. The reality was finally sheeting home: that ultimately growers could only grow and sell what companies would buy. Under the new system it was no longer company personnel but the growers' own representatives who were required to deliver the bad news that manufacturers were prepared to buy less and less New Zealand tobacco.

There was bad news, too, for Dick Stevens. Tremors within the federation saw him lose his position as chairman to John Hurley in 1976. Hurley had first challenged Stevens in 1972, with a measure of support but without success. But by 1976 the feeling had strengthened among the younger members of the federation that changing times called for a change of attitude and tactics. To this new breed of tobacco grower, it seemed the decades of adversarial tactics had failed to achieve progress or even stability. It was time to work together and use the board as a means of exchanging views rather than blows. After ten years as chairman, and almost forty years in the industry, Dick Stevens was on the federation sidelines, a position he did not relish. He challenged Hurley for the presidency again in 1977, saying he was as clear minded and capable of carrying on as ever, but the majority decision went to Hurley.[44] Stevens declined to accept nomination as vice-president. He retained his position on the Tobacco Board until 1979, when he was unseated by J.L. (Jack) Inglis in a three-way ballot.

The new chairman, John Hurley, was the youngest ever to head the federation. Even so, he had been a member of the Tobacco Board for ten years and had served on the executive since 1964. He came to the position of chairman with the idea that the whole executive should be more involved in decision-making. Meetings were more frequent and discussions more wide-ranging than under previous chairmen. Described as a 'big, iron-willed man with a fiery temper',[45] Hurley was certainly deeply commited to the industry. He reacted strongly to the series of events which unfolded during his presidency, becoming increasingly frustrated with the battle against outside forces.

The federation also changed its secretary during the 1970s. Willis Bond resigned in 1972 and was replaced by R.R. (Roy) Rance. Like his predecessor, the new secretary was also trained in Britain. A thorough worker, Rance immersed himself in federation concerns and quickly became almost indispensable. He was described as a gentleman but with a comical streak — certainly a sense of humour would be essential as he faced the monumental workload of the quota committee. The committee was under the auspices of the Tobacco Board, but almost all of the work was done in Motueka, with Roy Rance carrying the bulk of the administrative and calculation work. Later, Jeanne Rance was also employed by the federation as secretary-typist. Her efficiency and accuracy proved a valuable asset as she worked in partnership with her husband. The couple ensured the federation of complete discretion and a high degree of competence, both in handling quota work and in preparing price applications.

Delays in deciding and announcing the annual price of leaf were becoming longer and longer. The new system of pricing required the Department of Trade and Industry to examine the costs of production and related financial issues before recommending a price to the Minister. As audited costs were not available until some months after the season in question, it became customary to wait for these figures and for the yield per acre before establishing a price. While this was theoretically a fairer system for growers, it was also frustrating. It was often long after the crop had been bought before growers knew what they would finally receive. Small wonder that many growers lost interest in the relative value of specific grades as this was unknown at the time the tobacco was submitted for sale. The introduction of a grade-price schedule was intended to overcome this problem by providing a standard guide against which tobacco could be purchased. This allowed growers to see on the buying floor which grade their tobacco fitted into and what the likely price range was. Some doubted that the schedule improved the quality of domestic leaf – it did make it more expensive.[46]

The independent classification system envisaged in the 1974 Act became, in fact, panel assessment using classifiers drawn from company personnel. Despite the desire for independence from company influence, in reality only companies could provide people with the relevant experience. The panel worked on both buying floors in Motueka, on a roster system with John Reid of Wills as the initial coordinator. Reid reported in July 1975, after the first buying season under the new scheme, that 'five very differing persons' with different backgrounds had found themselves working together and it had taken some time to find common ground. The classifiers were distinguished by their uniform of gold-coloured coats, and the appearance of Wills classifiers on the Rothmans buying floor and vice versa gave at least a psychological appearance of impartiality. If the scheme opened a new era in the industry, Reid claimed there would none more delighted 'than the five men who make up the industry's first classification panel'.[47]

The board did begin the process of forming a completely independent panel with the appointment of Boyd Everett as a trainee classifier in the 1976-77 season. Working with companies and the federation, Everett's duties came to include measuring the acreage, advising the quota committee and generally undertaking a far wider range of tasks than classification. As the future of the industry became more and more uncertain, no further trainees were appointed by the board. When Everett resigned in 1984 his duties were taken over by Eugen Werthmuller, a retired Rothmans field officer and company buyer.

The late 1970s became a political battleground for the industry. Despite the intention to create better relationships, inevitable political and economic forces were lining up against the tobacco-growing industry. The oil shocks of the 1970s increased the cost of curing for the majority of growers: though a change to coal-fired kilns was encouraged, expense was a major problem. Inflation was affecting both growers and manufacturers. Companies faced international pressure to keep their costs down and the option of cheaper leaf from low-cost countries became more attractive as New Zealand costs rose.

The change of government in 1975 did not bring growers great hopes of continued support. The new Minister, Lance Adams-Schneider, soon gained a reputation as a straight-talker. Much of what he said was unpalatable to growers. Addressing his first federation AGM in 1978, Adams-Schneider told his audience the realities of their situation as he saw them. Recognising that growers were concerned with the falling domestic order, Adams-Schneider gave no hope of government intervention to reverse this trend. Neither could the government increase the tariff protection. Its stated intention was to gain better access to foreign markets for our butter, cheese, meat, wool and other products. In addition, the Minister warned growers that ever-rising prices for New Zealand tobacco made it 'increasingly tempting for manufacturers to use more imported leaf'.[48] The only bright note, although growers may have disputed this, was the Minister's intention to encourage research into alternative crops which could give comparable returns.

As if to reinforce the Minister's unwelcome message, growers again faced quota reductions in the 1978-79 season. In addition, it was suspected that Rothmans was dumping large quantities of cheap Australian leaf in New Zealand.[49] As in previous decades, the desperate growers grasped at any possible reason for the continued pressure on their livelihood. A meeting of Riwaka growers resolved to wage 'unrelenting war' on Rothmans. They urged the government to prevent the company from 'dictating policy'[50] by controlling the domestic industry while importing large quantities of cheap, low-grade leaf. The New Zealand Workers' Union weighed in on the side of growers, seeing quota cuts as a threat to their members employed in seasonal work. Their local representative, J.A.H. (John) Krammer, believed the industry was 'the victim of political deceit'. After previous reassurances from the Minister, the local National candidate, Ruth Richardson, and the branch chairman, T.E. (Tom) Inglis, growers now heard 'a strange silence from those quarters'. History showed that the local industry had always been at the mercy of tobacco companies, Krammer stated, and now all in the district should unite to protect their industry from manipulation 'by multinationals, politicians and fellow travellers'.[51]

Local grower D.I. (Dennis) Crisp set out the growers' case in a long letter to the *Nelson Evening Mail*. Acknowledging that tobacco growers were 'farming a commodity which ... is deemed by many to be harmful', Crisp asserted, nevertheless, that 'no government would be stupid enough ... to ban the use of smoking products'. Crisp could therefore see little reason to demolish the local industry and encourage growers into alternative crops. He could foresee a day when those who opted for kiwifruit and found it unprofitable could find themselves encouraged by a future government to take up tobacco farming. 'After all, they could accurately argue that we have the right soil and climatic conditions, that we'd save the country about $12 million a year and create employment for about 1500 people.'[52]

Manufacturers had fared somewhat better under the new government. Affected by liquidity problems in the mid-1970s, companies approached the government to provide backing finance to enable them to purchase the domestic crop. Without such finance, it was claimed, there was a possibility that no New Zealand tobacco could be purchased. Adams-Schneider introduced the necessary legislation to enable the government to provide this finance in 1976.[53] This assistance was then accepted by manufacturers on a continuing basis, with repayment being spread over eighteen months. The scheme operated through a Tobacco Board overdraft facility with the Reserve Bank and by 1979 this facility involved a sum of $13 million. Bill Rowling claimed that the finance provided by the government to underwrite the purchase of the domestic crop had carried an implicit understanding that companies would use more local leaf.[54] Rowling complained bitterly that companies were not meeting their part of the bargain. Although the mandatory domestic usage had been retained at 30 per cent, the voluntary component of domestic usage was being steadily eroded. Restrained by international trade pressures, the government took no action to reverse this trend. Adams-Schneider, in essence, left the question of greater use of New Zealand leaf up to growers, and told them that even maintaining the present level of usage might depend on becoming more competitive with imported leaf.[55]

The situation took a more dramatic turn in 1979 when the board experienced difficulty in arranging the necessary finance for the crop. Companies threatened that without this finance there would be no domestic order.[56] Frantic negotiations took place between new board chairman, E.S. (Eric) Highet, and Cabinet officials – finally the finance was granted and the order placed. The clear signal, however, was that the government and Treasury were becoming restive over the arrangement and were seeking a way to end it.

Meanwhile the quota committee struggled on with its mammoth and divisive task. Overproduction became a problem for both growers and companies. Stocks of domestic leaf were again higher than manufacturers felt they needed and the order was reduced even further. In 1979 the domestic order amounted to a 16 per cent cut in quotas and companies also proposed to offset over-quota leaf against the following year's order. Thus, a good season would see growers penalised by an even lower basic order the following year. The quota hold-over became a contentious point, as Bill Rowling pointed out that nothing in the 1974 Act allowed this action. Ever active on growers' behalf, he described the quota cuts of 1979 as 'bureaucratic lunacy'.[57] In October 1979, after taking legal advice, the board rescinded the motion which had implemented the policy.[58] The long-standing growers' call for a 'smoothing' scheme, whereby excess production could be offset against poor seasons, was revived but no agreement was reached. A compromise was eventually arrived at, with growers able to spread over-quota leaf over the following three seasons without having their basic quota reduced. In the event, growers agreed to a later proposal to have their over-quota leaf from the 1978-79 crop accepted as the first purchase of the following buying season, leaving their basic quota intact. The smoothing scheme never eventuated.

The situation in 1979 was viewed with concern by the New Zealand Workers' Union, whose local organiser, John Krammer, formulated plans for industrial action to prevent the importation of raw leaf. Krammer was convinced that the local industry, and therefore his members' employment, was under threat from multinationals and lack of government support. On a trip overseas he contacted union organisations across the US and in Britain, as well as European and Soviet satellite countries, seeking international backing for a ban on imports by New Zealand tobacco companies. Krammer believed that such a ban would force high-level discussions between all parties and would strengthen the growers' hand.[59]

This type of action was supported in a plan submitted by grower Ian Clouston, who advocated joining forces with the Federation of Labour to create nationwide support for the growers' case. Clouston's main point was an increase in domestic usage. Focussing on the low usage by Rothmans, he suggested direct action against the firm's operations at Napier, including a load-out ban or a drivers' strike. Clouston's attitude was militant and he urged the federation to be prepared for the consequences of direct action. Growers must be ready to contribute to a strike fund if necessary and, even more aggressively, he believed growers 'must be prepared to destroy Rothmans'[60] if the company was not prepared to negotiate. Clouston put forward four options, starting with the most desirable position for growers. His bottom line was a requirement for companies to purchase all leaf from a set acreage and for imports to be the variable factor in the local manufacturing industry.

The federation's strategy committee, set up to deal with the crisis, did not agree with the militancy of these proposals. They did, however, formulate an 'agreement' to put to manufacturers which contained some elements of Clouston's plan. The proposed agreement, which was later modified into a 'letter of intent', called for a recognition of the domestic industry and an undertaking to purchase all quota-grade leaf from a specific area.[61] Initial approaches were made to W.D. & W.O. Wills, to test its response to the proposal. Wills received the delegation courteously and discussions were open and frank, but the company declined to sign the letter of intent, giving reasons which were mainly financial.[62] Some of the impact of the initiative was lost when the delegation found that Wills already had copies of the letter and related documents. A strategy committee member, Dennis Crisp, confirmed he had discussed almost all of the committee's proposals with Wills personnel, thereby sparking severe criticism from other members. Krammer's plans were similarly undermined by personal disputes within the union.

The strategy committee eventually disintegrated as members struggled with the complexity of their problems and the difficulty of forming a focused approach. Three of the Ward Three delegates resigned in frustration. Mervyn Heath expressed the view that to 'send Strategy Committee members out into the wilderness' to solve the entire industry's problems within a week was 'a little optimistic'.[63] The problems remained largely unsolved and growers remained unsettled and uncertain over their future.

The new Tobacco Board chairman, Eric Highet, appointed in 1978, was finding the situation just as difficult. Highet was the first chairman to be appointed on the nomination of the federation – with a background in farming services he was seen as much closer to understanding the specific concerns of growers. Having studied the 1974 Act closely, Highet began his task with hopes of good relationships on the board, despite the anomalies he felt were inherent in the legislation. At his first meeting, however, he was required to use his casting vote.[64] This was an early, and somewhat unwelcome, indication that the task of board chairman would be far more than the half-day-per-week mediating role he had expected.

Resident in Nelson, Highet found much of his time now consumed by trips to Wellington to deal with the growing tension as companies addressed their respective situations. Phillip Morris, claiming a shrinking market share, sought to place a nil order for several seasons in a row and the other companies refused to take up the extra quota. In 1979 the board declined the Phillip Morris request and decided to enforce the provisions of the Act, which required companies to take up their firm orders. Even so, members of the board were aware of considerable difficulty in implementing the new Act. Definitions were causing trouble, as was the lack of specifics with regard to the quota system. Both John Hurley and Gerald Hunt now publicly stated that the board was not pulling together as a team for the good of the industry. Introducing Adams-Schneider to the 1979 AGM, Hurley told the gathering what they already knew – that their industry was a 'continuing story of crisis'.[65]

By the end of the 1970s the clouds were thickening over the Motueka growers. Late in 1979 Adams-Schneider announced a departmental investigation into the local growing industry with a view to restructuring and streamlining what he now considered to be an inefficient industry. The problem of finding the annual finance was also a factor in his analysis. The departmental team, consisting of officials from Treasury, the Ministry of Agriculture and Fisheries, and the Ministry of Trade and Industry, toured the district in late 1979, and presented its report in early 1980. This was followed almost immediately by a government 'package to restructure the growing industry. Growers were shocked by the ferocity of the proposals, which attacked the very foundations of their livelihood, and by the speed with which the programme was carried out.

NOTES
1. Interview with John Husheer, 9.12.96.
2. Tobacco Growers' Federation minutes, 6.3.70.
3. Ibid, 3.7.70.
4. Ibid, 18.11.70.
5. Ibid.
6. Ibid, 3.7.70.
7. Ibid, 26.11.70.
8. Ibid, 26.11.70.
9. Ibid, 3.4.70.
10. Ibid, 9.2.71.
11. Ibid, 28.5.71.
12. Ibid, 3.12.71.
13. Ibid, 20.2.70.
14. Tobacco Growers' Federation submissions to the Minister of Industries and Commerce, Warren Freer, 1973, federation files.
15. Parliamentary Debates, 1971, p.2446.
16. Ibid, p.2447.
17. Ibid, p.2449.
18. Ibid, p.2451.
19. Tobacco Growers' Federation minutes, 13.10.71.
20. Ibid, 24.8.71.
21. Ibid, 4.8.72.
22. Ibid, 24.8.71.
23. Ibid, 10.1.73.

24 Ibid, 8.1.73.
25 Ibid, 17.4.73.
26 Tobacco Board report, 1971, p.8.
27 Interview with R.W. (Rob) James, Motueka, 29.8.96.
28 Ibid.
29 Tobacco Growers' Federation minutes (rough hand notes of secretary), 15.12.77.
30 Letter from P.D. Connery, 21.12.77, Tobacco Growers' Federation files.
31 Tobacco Growers' Federation minutes, 15.12.77.
32 Ibid.
33 Letters from Tobacco Growers' Federation to three growers, 3.5.78, federation files.
34 Bulk curing eliminated the need to tie tobacco onto sticks. Instead the leaf was pressed into clamps – metal frames fitted with spikes to hold the loose leaf firmly – and cured in bulk.
35 Parliamentary Debates, 1973, p.852.
36 Supplementary tobacco manufacturers' submission to the select committee, 1973, p.1, federation files.
37 Initial manufacturers' submission to select committee, 1973, p.9, federation files.
38 Tobacco Growers' Federation submission to select committee, 1973, pp.3 and 16, also supplementary submission pp.3 and 12.
39 Interview with former Rothmans manager in Motueka, R.E. (Rollo) Wilkinson, Motueka, 21.8.96.
40 Interview with Gerald Hunt, 30.8.96.
41 Tobacco Board report, 1977.
42 Supplementary tobacco manufacturers' submission to the select committee, 1973, p.4, federation files.
43 *Marketplace*, Vol. 1, No. 1, 1972.
44 Tobacco Growers' Federation minutes (rough hand notes of secretary) of executive meeting, 5.10.77.
45 *Nelson Evening Mail*, 1.8.81.
46 Tillson, G.B. (Geoff), History of Tobacco Growing in New Zealand, unpublished paper, c. 1980, p.15.
47 Article by John Reid, June 1975, Tobacco Growers' Federation files.
48 Adams-Schneider's speech notes for 1976 Tobacco Growers' Federation AGM, federation files.
49 *Nelson Evening Mail*, 8.12.78.
50 Ibid, 8.12.78.
51 *Motueka News*, 13.12.78.
52 *Nelson Evening Mail*, 6.12.78.
53 From Rowling's speech to 1978 Tobacco Growers' Federation AGM, federation minutes, 22.9.78.
54 *Nelson Evening Mail*, 23.9.78.
55 Ibid, 26.9.79.
56 Interview with former Tobacco Board chairman E.S. (Eric) Highet, Nelson, 27.9.96.
57 *Nelson Evening Mail*, 27.9.79.
58 Ibid, 24.10.79.
59 Paper supplied by J.A.H. Krammer, October 1996.
60 Paper submitted to Tobacco Growers' Federation strategy committee by I.D. Clouston, federation files.
61 Letter of intent prepared by strategy committee of Tobacco Growers' Federation, 1980, federation files.
62 Minutes of strategy committee, Tobacco Growers' Federation, 15.4.80.
63 Letter from M. Heath to strategy committee, Tobacco Growers' Federation, 22.4.80.
64 Interview with Eric Highet, 27.9.96.
65 *Nelson Evening Mail*, 28.7.79.

CHAPTER SEVENTEEN

Savage Surgery

The 1980s began in turmoil and ended with a shattering blow. Even before the inter-departmental study group's report was presented, the Prime Minister, Robert Muldoon, announced the government's intention to restructure the tobacco-growing industry and remove 'inefficient' growers. The outline of the plan released by Adams-Schneider in July 1980 included phasing out mandatory domestic usage and abolishing licences and quotas. It removed the price-fixing system and provided for the gradual dismantling of import restrictions on cigarettes and manufactured tobacco. Reserve Bank funding of the industry would be wound down over four years. The requirement for companies to notify their forward orders would also be discarded – in effect, the industry would be completely decontrolled by 1985.

Shockwaves reverberated around the district as growers' worst fears appeared about to come true. All the protections they had struggled for since the 1930s would disappear in one sweeping onslaught. John Hurley reported that the immediate effect of the government's announcement was that some companies had not placed their forward orders before the board. They were, therefore, breaching the 1974 Act, which still governed the industry.[1] Growers found references to inefficient production particularly galling, as the average yield per hectare of 2090 kilograms was considered high in world terms. Criticism was aimed mainly at the inter-departmental study group. The group's report would not be made known to growers until 1984 but this did not inhibit criticism. An article in the agricultural journal *Straight Furrow* probably spoke for most growers when it described the group as 'a collection of bloody ning-nongs who didn't know anything about the industry'.[2]

After a somewhat bemused initial reaction, the federation swung into attack mode as it had so often in the past. Emergency meetings and urgent deputations to the Minister were arranged, extensive submissions were prepared and growers were kept informed as events unfolded. Once again the local community was mobilised in an effort to protect the industry which provided such an injection into the district economy. A public meeting was called. More than 800 people arrived at the Memorial Hall in Motueka to hear federation speakers and the local mayor, David Kennedy, spell out the possible effects of the government's proposals. Loud-speakers were set up for those unable to get into the hall. Mayor Kennedy read a metre-long telegram from Lance Adams-Schneider, the sight of which brought a degree of laughter from the audience. Further laughter greeted the Minister's statement that the details of the package could not yet be revealed as 'it was still being fine-tuned'.[3]

The meeting was preceded by a parade of tobacco machinery in High Street. Many shops closed for the parade and remained closed during the meeting. Hotel patrons spilled onto the streets to watch the strange procession of machinery which had seldom been seen off the farms before. Locals were well aware of the impact any cuts in tobacco growing would have on their area. Less tobacco grown meant fewer seasonal jobs, less money circulating in the town and less overall prosperity for the whole district. The New Zealand Workers' Union again lent its support. Union representative Trevor Wearne told the meeting that the union was 'not going to sit idly by and watch the government kill an entire industry and virtually dismantle the Motueka community'.[4] Mayor Kennedy called for a Federation of Labour ban on handling imported tobacco and questioned the government's right to label growers inefficient when state operations such as Air New Zealand and the railways ran at a loss.

Bill Rowling was equally scathing, calling the plans 'a sellout'. The industry provided millions of dollars in tax and excise revenue as well as jobs and business opportunities – the community must unite to protect its historic industry. Tobacco industry leaders addressed the meeting, calling on community support to prevent the destruction of the economic viability of Motueka. Leaders of other horticultural industries and business organisations spoke in support of the growers' case for a more sympathetic deal from the government. A third-generation tobacco grower, E.W. (Ernie) Drummond, moved a resolution to set up an action committee designed to impress on the government the strength of feeling against the package. The meeting supported the move unanimously.

Opposition was not restricted to growers, unions and opponents of the government. The National Party dominion councillor for the Tasman electorate, Tom Inglis, saw the proposals as a serious threat to Motueka and its people. 'It isn't just a question of pressing the calculating button [and saying] well, some of you people are going to the wall'.[5] Also a third-generation tobacco farmer, Inglis felt growers had an obligation to the district and to workers, many of whom had made financial commitments in order to stay in the district and be available for work. Inglis's sentiments were endorsed by other local party officials who voiced their displeasure at the party's annual conference in Auckland. At this conference Tom Inglis criticised Muldoon for announcing that tobacco growers 'could grow flowers instead'.[6] The *Nelson Evening Mail* found it 'remarkable' that the government approach to restructuring was so clumsy and insensitive and that the lack of assurance on the future stability of the district was leaving the 'township of Motueka feeling exposed, nervous and bitter'.[7] The *Mail* asked whether the phasing out of tobacco, which saved millions of dollars in overseas funds, would be replaced by export crops. If so, what were these crops, what evidence was there that they would be more profitable to the district, and what effect would an increase in numbers growing them have on existing farmers? The *Mail* found the whole process an exercise in cost accounting with little regard for human concerns. The government had exhibited a 'callousness to be regretted'.

The uncertainty of the future quickly affected Motueka. As early as mid-July 1980 a local engineering firm, J.E. Trevett New Zealand Ltd, had been forced to lay off four of its seven workers. Another firm announced that more than $200,000 worth of contracts had been cancelled following the announcement of the government's plans.[8] By August firms supplying calico and plastic for seed-bed covers, sprays and building services also reported cancellations.[9] Retailers, too, were beginning to feel nervous of their future. The Tobacco Board again appeared powerless. Ignored by the government and its officials, the board had no power to influence either the Minister or the manufacturers. John Hurley described the July meeting of the board as being 'like a gathering of funeral directors'.[10] To add to the general gloom, manufacturers announced a reduced order for the coming season, 300,000 kilograms down on the previous year. Taking into account the over-quota already held by growers, this was virtually a cut of a third in the total crop and represented an area reduction of 800 hectares in three years.

Growers' representations on the severe impact of the proposed restructuring on growers and their community were largely dismissed or disregarded by government officials. A comprehensive plan to restructure the industry without disrupting the local economy, while still saving the government money, had already been presented to the Minister. The action committee pressed this idea, which involved the formation of a growers' cooperative to retain ownership of tobacco leaf during the two-year maturation period. This period was currently covered by the Reserve Bank overdraft facility. The plan also proposed an incentive rebate on tariff for manufacturers if they used a greater proportion of domestic leaf. The action committee met with Adams-Schneider and his officials at the end of July 1980, with further proposals. These included a new tariff structure, the elimination of the district average price, measures to encourage raised standards, and a producer board or other authority to contract for, and market, the domestic crop. Bill Rowling, who accompanied the delegation, stated that, while a decision could not be expected immediately, he now hoped that in-depth discussions would take place 'before hassling an industry ... so critical to a community like Motueka'.[11]

As growers got on with preparations for the coming season, attention turned to the role of companies in the stability of the growing industry. In a rare outburst against manufacturers, the incoming federation president, Gerald Hunt, claimed the industry would never progress until there was a government 'strong enough to withstand the pressure of a multinational'.[12] This was a reference to Rothmans, which was thought to be pressuring domestic usage downwards. Hunt's outburst was even more surprising in the context of his close ties with the National Party. The 1980 federation AGM had a distinctly militant tone. Growers expressed their discontent in resolutions censuring Rothmans for a 'policy of deliberately and systematically setting out to destroy the domestic tobacco-growing industry'.[13] Attacks were also made on government officials for their lack of knowledge of the industry. Gerald Hunt claimed that the officials concerned did not understand the industry they had set out to restructure. They had been amazed, when visiting the area, to see how many of their proposals were already in place. The personal effect on growers was also illustrated. One grower said he had previously employed eighteen workers but his quota was now so small he was 'down to two part-timers, with my wife and I doing much of the work'.[14]

Speaking at the same meeting, union representative Trevor Wearne told growers the industry should make its position a public issue and should let the whole country know what was happening. Bill Rowling, attending his seventeenth federation AGM, said the package had been presented by 'those who don't know tobacco from dock leaves'.[15] He reminded growers that the government received $142 million from tobacco revenue – this would not change even if there were no domestic industry. Rowling had been equally scathing in Parliament, describing the proposals as a package 'put together by some very inexpert so-called experts who perambulated around the district some months ago'.[16]

In October 1980 the industry received a reprieve. Adams-Schneider announced that he was referring the question of restructuring to the Industries Development Commission (IDC). The Minister told the Tobacco Board he had no wish to bulldoze through his package if further study and discussion would result in measures acceptable to all parties. The move was welcomed as a necessary slow-down in the headlong race towards restructuring and would buy time for further submissions and negotiations. Amongst these negotiations was the suggestion of financial compensation for growers willing to leave the industry – figures ranged between $6000 and $8800 per hectare.

Companies also took the opportunity to lobby the Minister, with Laurie Doolan, of Rothmans, putting his company's case privately in November 1980. Doolan pointed out that the board had been far from unanimous in its eventual submission to the Minister – indeed, his company strongly disagreed with parts of the board's recommendations. His main dispute was with the proposal for the board to retain power to deal with over-quota leaf after the 1981-82 season. Doolan believed that over-quota tobacco should be left at the risk of the grower and that since such leaf was cheaper for the grower to produce it could be sold at lower prices.[17] Growers, on the other hand, maintained that over-quota leaf was grown under licence on licensed areas and therefore represented efficiency.

The board's submissions went forward without alteration, along with strong representations from the federation. By now the new crop was planted out as growers had always to continue their seasonal round of activities whatever the uncertainties. As they had for decades, women crouched at tobacco beds through to mid-November, pulling plants in the early summer heat, often buffeted by the strong winds which characterised planting time. Acreage had been greatly reduced but there remained an impressive area under cultivation. The freshly worked, neatly arranged tobacco fields were an annual novelty for visitors and a relief for growers as their planting was completed.

In December 1980, however, the Minister reaffirmed his plans for the industry, having effectively disregarded industry and community suggestions. Restructuring plans were much the same as before, softened only slightly by the offer of compensation of $7000 per hectare for growers leaving the industry for other crops. The proposals now also included the removal of price control on cigarettes and manufactured tobacco products. The growing industry was in confusion. None of their own alternative suggestions had been incorporated in the package and no provision was to be made for a marketing authority or producer board. Given the insecurity and uncertainty, significant numbers were expected to accept the compensation offered. The federation warned of the danger of losing community support if growers appeared likely to bale out in large numbers. The executive itself was sharply divided on the issue. The absolute defence of tobacco growing appeared to be diminishing as alternatives were considered. K.J. (Kevin) Fry illustrated this dilemma with his statement that 'perhaps all growers should take the compensation if the Package is the best deal the Industry will get'. John Hurley, increasingly incensed by the situation, declared he would refuse to take any part in 'picking up the pieces of an Industry destroyed... by officials'.[18] The executive appeared as confused and divided as the rest of its members.

Meanwhile federation officers were summoned to Wellington for 'consultation'. They were confronted by the Minister and officials from Treasury, who seemed uninterested in any of the growers' arguments. The government's determination to rid itself of the annual financing of the industry and its ongoing internal troubles seemed absolute and unchangeable. The Minister had sidestepped his earlier decision to wait for the findings of the IDC. He now requested growers and manufacturers to prepare an agreement in principle to cover the transition to a free-market environment. The visiting delegation was by no means satisfied. In response to a question regarding displaced workers, officials declared these would be taken up by diversification – no evidence was offered to support this statement. Some heat was generated when a Treasury official, Roger Kerr, suggested that departmental officials had 'come up with a high-quality decision'. John Hurley retorted scornfully that it was 'anything *but* a quality decision – he would call it a scandalous one'.[19] Highly dissatisfied, but having made no impression on the Minister, the delegates returned to Motueka with their hopes battered and the task of relating the news to their fellow growers.

A special meeting of growers was called immediately but there was little the executive could say to placate its members. Repeated letters and delegations to the Minister had achieved almost nothing. Eric Highet found some cause for hope in the proposal for the board to formulate an agreement on transitional measures. It was possible that such an agreement between growers and manufacturers would force amendments to the package. Growers were again conscious of their lack of influence and the way in which their case had been ignored by government. A list of grievances was expressed succinctly by Riwaka grower David Cederman. He could see no reason for any restructuring other than to remove the $13 million lent to the manufacturers, but thought it might be a political move against Bill Rowling. Cederman alleged that the quota committee had allowed a lot of 'playing around' with the leasing of quota, which he found disgusting. The $7000 compensation was offered to 'buy growers' mouths and to keep them quiet'.[20] Finally, he made an appeal for growers to be taken into the confidence of the leaders of the industry.

Again growers were divided and confused. Some were becoming resigned – others wished to maintain a militant stance, calling for meetings around the country. For the first time a woman's name is recorded in federation minutes. Elizabeth Bartlett urged growers to plan to work together to assure their future. After much debate the meeting resolved to continue to fight the government package. Chairman Gerald Hunt then asked for some breathing space to deal with the present difficulties. There was to be little such breathing space. At the end of December 1980 Rothmans advised the board of its minimum forward order for 1981-82. The company would require only

250,000 kilograms of domestic flue-cured tobacco, down from 1,880,000 kilograms in 1980-81. The given reason was the need to absorb the quantity of over-quota leaf already in the system – the company acknowledged that this projection would be 'disappointing'.[21]

Growers generally were probably unaware of these company moves, being by now totally absorbed in the harvest season. The summer of 1980-81 was particularly dry. Continual irrigation was necessary, with some crops showing drought symptoms until heavy rains late in February brought a steady improvement in leaf quality. Meanwhile discussions on the government's package continued at board level. Eric Highet felt that, as officials had now been directed to listen to industry participants, the message should finally get through. Manufacturers were showing a willingness to negotiate, provided their own position was secured, he said.[22] Not all were as optimistic. At a public meeting called by the New Zealand Workers' Union Bill Rowling warned that a nil order from companies was possible if surplus stock was used as a weapon against growers. Both Rowling and the Motueka mayor, David Kennedy, advocated a ten-year adjustment period rather than the proposed phase-down over five years.

The surplus-stock question was put into context by Gerald Hunt who attributed it solely to the reduction in usage of New Zealand tobacco. Actual production had dropped steadily but company usage had been cut more quickly. As demand for tobacco products reduced it was domestic leaf which was affected because manufacturers adhered to their established blend ratios. Union speakers were more direct in their criticisms, both of companies and of the government. A Federation of Labour economist, Alf Kirk, spoke of the general trend towards restructuring and of some of the inherent weaknesses in this philosophy, including an increasing reliance on imported goods. The president of the Nelson Trades Council, Larry Sutherland, blamed pressure from multinational interests on a weak government, and described the restructuring as 'a complete farce'.[23] A woman seasonal worker commented that, while growers were set to receive $7000 per hectare for leaving the industry, there would be no such compensation for those about to lose their jobs.

The report of the Industries Development Commission was awaited with apprehension by growers: the commission's findings and recommendations were expected to reinforce the provisions of the government's package. There was speculation that the commission favoured the use of land in the Motueka area for export-earning crops rather than the import-substitution role of tobacco. Submissions to the commission were made by the federation, the board, manufacturers, and the Nelson-Bays United Council. The companies agreed with growers that the phase-out time was too short, recommending that at least eight years would be more in the industry's interests. The United Council was concerned with land use, making an 'impassioned plea' for a longer changeover time for those growers moving into alternative crops. The return from these alternatives was by no means proven, argued David Kennedy, and there was need for a viable tobacco industry to provide the cashflow and capital needed to finance any transfers.[24] The federation again stressed the unique position of tobacco growers. Facing stiff competition from subsistence-economy countries, growers also had to deal with internal inflation and rising costs. Being licensed and controlled by quotas, growers were not able to increase production to offset these factors and, with tobacco growing confined to a small area within one electorate, the industry had little national clout.

In the event, the commission's report was even more severe than expected and the government moved to modify the recommendations. The Tobacco Industry Amendment Act 1981, contained the elements of Adams-Schneider's plan and the commission's report, but with some movement on tariff protection and the introduction of a tariff rebate for manufacturers. The Act also offered a further review of the industry in 1984-85. Growers were not placated by this. Newspapers around the country waded into the argument. Wellington's *Evening Post* called the package a 'killer punch', a description that acting Minister of Trade and Industry Warren Cooper called misleading and mischievous.[25]

Companies were not unaffected by the package. While they favoured the removal of specific domestic usage and long-term forward orders, they were disturbed to see the crop financing arrangements being phased out. Adding to the confusion was the fact that the compensation payments, now called special assistance grants, were already being taken up by numbers of growers. Some 717 hectares had been withdrawn from production by June 1981, involving a payout of more than $5 million. Indications were that 65 per cent of tobacco land would go out of production before the next planting season. The assistance grants would ultimately cost the government a total of $9,600,000.

Some were beginning to accept the inevitability of the new situation but growers were by no means unanimous in this. John Hurley, no longer the federation chairman but still a grower representative on the Tobacco Board, railed bitterly against the government's proposals. In September 1981 Hurley tendered his resignation as a board member, informing the Minister that he 'could not possibly represent growers when their interests have been so undermined'.[26] Hurley was not replaced, leaving grower representation at two. With the withdrawal of Phillip Morris from buying domestic leaf in March 1981, manufacturer representation on the Tobacco Board had also been reduced to two. Legislation to adjust board numbers to seven was incorporated in the 1981 Act.

At the same time there was talk of a split in the federation, although nothing eventuated. Significantly, when the 1980 inter-departmental study group's report was finally received in 1985, it identified the divisions between growers as a cause of 'ferment' within the industry. 'Malcontents' had seized on rumours and misunderstandings 'to the overall disadvantage of the industry'.[27] The situation was uncannily reminiscent of the 1950s. The federation's 1981 AGM was a mixture of cautious acceptance and warnings of conflict. Some agreed that the quotas as allocated were generous for those remaining within the industry, while others saw difficulty in the Tobacco Board now attempting to serve three masters. It would not be possible to satisfy growers, manufacturers and the government, they claimed. Bill Rowling warned of the dangers of overproduction in alternative crops, especially kiwifruit. Other crops, such as berries, were 'only valuable when you have a market'.[28]

Despite all protestations and rhetoric, growers had little choice but to adjust to the new environment. Decisions had to be made, and a variety of options were taken up. Some took the opportunity to quit tobacco growing altogether, accepting the special assistance grant in order to diversify into other crops. Others sold their farms, with or without quota, and moved on to completely new fields. The remainder, now numbering 122 and producing from 470 hectares, concentrated on the business of planting for the coming season. The threat of a minuscule, or non-existent, company order had receded, with the domestic order now confirmed at 1,896,946 kilograms.[29] The danger now was that the area planted would be insufficient to give an 'economic throughput level' for company processing facilities. Almost immediately, moves began to allow some growers to 'buy back' into the industry by repaying their special assistance grants through an annual payment of $1750 per hectare. Company personnel in Motueka were active in this process, encouraging growers not only to re-enter the industry but to enlarge their present acreage.[30] By 1982-83 grower numbers had risen again to 135 and production area to 657 hectares. This trend continued until acreage reached 824 hectares in 1985-86.

Relations between the federation and manufacturers, however, improved markedly during this period. Federation chairman Gerald Hunt was influential in this. Hunt's natural inclination to find solutions rather than problems found a ready acceptance among company personnel. A personable, diplomatic man with an ability to express ideas well, Hunt was conversant with the internal politics of the industry. A board member since 1966, he was well known to company personnel and, as a former National Party candidate for Tasman, he also had useful political connections. Gerald Hunt differed sharply from some of his predecessors in that he believed in cooperation rather than confrontation. He would suffer for this as some growers viewed cooperation as 'selling out'. Others distrusted Hunt's political affiliations. The old suspicions and enmities died hard.

But genuine efforts to work together began to bear fruit. Agreement was reached in July 1983 on the major points under which the industry would operate. Companies agreed to purchase leaf only from federation members and to place minimum forward orders for three years. Independent classification would continue, as would the provision for the purchase of a small percentage of over-run tobacco. The annual price would be negotiated directly within the limits of competitiveness with imported tobacco, and companies agreed that the price would never decline. The government favoured this type of industry agreement rather than a marketing authority or a return to the provisions of the 1974 Act. The federation thought it had done well considering the fears expressed when restructuring was announced.[31]

As early as 1981 the federation was invited to join the Tobacco Institute, especially in the fight to nullify the work of the anti-smoking lobby. In 1983 Laurie Doolan of Rothmans attended his first federation AGM. David Weaver of Wills was also present at this meeting, and both told growers they owed a great debt to their present chairman, Gerald Hunt, for the way he had worked to stabilise the industry. Under Hunt's chairmanship industry relationships had improved and dialogue between growers and manufacturers appeared more constructive. A public relations meeting held by Wills for its growers in 1984 carried an air of euphoria and optimism. The company officers present, David Weaver and Stuart Watterton, spoke convincingly on the company's plans and reminded growers of recent capital improvements to the local factory, estimated to have cost up to $350,000.[32] Wills growers, at least, were reassured that their industry was coming out of the uncertain and fragile environment created by the restructuring process. Bill Rowling, reflecting on his twenty-year involvement with growers, was less positive. With the legislative slate about to be wiped clean, Rowling was disappointed and 'somewhat fearful' about the future. The 1974 Act, born out of forty years' experience and many battles, was now a 'smouldering heap of ruins'. He could only hope there were not more trials and tribulations ahead.[33]

Despite these fears, the growing industry pressed on, with the federation now more corporate in its approach. Moves to streamline the executive were made in 1983 with a reduction to six members, directly elected, and the abolition of the ward system. The 143 growers remaining in the industry appeared to be coping well with the changes as the National government was defeated in the 1984 general election. The incoming Labour government, however, changed nothing for tobacco growers. The long-time support of Bill Rowling was no longer available, and restructuring and deregulation fitted neatly into the prevailing philosophy of the new Labour Cabinet. The new Minister of Trade and Industry, David Caygill, listened attentively, and apparently sympathetically, to growers' submissions and then proceeded with the previous administration's programme. The industry would be finally deregulated in 1986. It would be in the same position as before the 1935 legislation, or perhaps a worse one, since there was little indication that any company was specifically committed to the New Zealand industry and no indication at all that the government supported growers.

While no change in policy direction occurred for growers under the new government, the devaluation of the New Zealand currency immediately after the election gave back domestic tobacco its competitiveness with imports and secured the level of production for a time. Alternative crops were beginning to appear viable, although the claim that a living could be made from one hectare of kiwifruit was viewed with justifiable scepticism. Just as this mood of cautious optimism began growers were faced with the 1984-85 review they had been promised. Again the work of departmental officials gave little cause for celebration. The report restated the severe recommendations of the 1981 IDC report, and gave growers no hope that the official direction of the past five years would be reversed.

The Ministry of Agriculture and Fisheries representative on the Tobacco Board, Moore Baumgart, was incensed. He told growers they had been knocked about by 'special people' when what the industry needed was fair treatment.[34] In the event, the report was thoroughly discredited by submissions from industry participants, and growers retained their footing on the slippery slope for the moment. At the same time, however, key figures were leaving the scene. Bill Rowling retired from Parliament in 1984, Rob James left the research station in the same year, and Eric Highet resigned as board chairman in 1985. Wills assured growers at their 1985 AGM that the company intended to remain in the area, but there was little reassurance in the words of David Caygill at the same meeting. New Zealand tobacco growers had been protected for forty years, stated Caygill, and this had led to the problems of oversupply. Government measures were designed to create a viable industry from one which had been large and inefficient prior to 1980. It was not government policy to increase protection but it would not be fair if tobacco was ahead of other sectors in the removal of protection. This minimal assurance was accompanied by an indication that the government would attempt to have New Zealand leaf qualify as Australian domestic leaf under the Closer Economic Relations Agreement (CER).[35] The implications of the withdrawal of legislative support were becoming clear.

The federation itself suffered a severe blow early in 1986 with the sudden death of the secretary, Roy Rance. Rance's thorough knowledge of the industry and his ability to work on growers' behalf, formulating policy papers and preparing submissions, had been a priceless asset through the difficulties of the early 1980s. The federation now faced the daunting prospect of replacing Rance; Jeanne Rance had also decided not continue working after the death of her husband. In the new environment the federation executive was expected to function more as a board of directors, and professional backup of the type Rance had provided was urgently needed. By mid-1986 the federation had appointed M. D. (Mel) Cox as its new secretary. The American ex-ship's purser quickly settled into the complexities of his new job. Mel Cox's friendly, unruffled personality was soon accepted by growers and he quickly gained expertise with computers, which would prove invaluable to the federation. Cox's task, as the industry faced the second half of the decade, became a financial and legal labyrinth.

All legislation affecting the growing of tobacco was repealed in 1986. The administration of the Crop Insurance Fund, now valued at just over $1 million, became the responsibility of the federation. In the decontrolled market, negotiations on the domestic order and price would be directly between growers and manufacturers. This latter created an almost immediate problem as, although the federation was the logical vehicle to administer quota and negotiate prices, under the Commerce Act this was seen to be a monopoly situation. Urgent action was required. The federation applied to the Commerce Commission for a restrictive trade practice departure. In the meantime growers were enabled to sign individual contracts with companies. As predicted, the situation was uncannily reminiscent of pre-1935 days.

The federation continued its attempts to maximise the situation. Plans for a marketing authority had not been abandoned but it was proving difficult to obtain official approval. Again, the Commerce Commission objected to the possibility of a monopoly organisation. Federation officers then began moves towards forming a cooperative to act on behalf of growers. When Ernie Drummond became president of the federation in 1987, the formation of a cooperative was his main objective. A practical, enterprising man, Drummond had served on the federation executive since 1977. He had been deeply involved in the Agricultural Training Council, managing to bring tobacco trainees under the scheme. For the first time a structured and approved training course in tobacco cultivation was made available to young prospective growers. On his retirement as federation president, Gerald Hunt expressed the belief that 'just as long as people continue to smoke, the tobacco-growing industry will be needed in New Zealand'.[36] He was confident that the federation would be ably led by his replacement. Hunt was only partly right. Ernie Drummond's foresight and commitment to his industry could never be questioned, but he was to head the federation through one of the most difficult and disappointing periods in its history.

Companies continued to prune the domestic requirement, placing an order for 1986-87 which required a six per cent cut in basic quotas. As growers struggled with the continual shrinking of their industry, the local MP, Ken Shirley, struggled to explain his government's policies to growers at their 1986 AGM. Battling resentment over rumours that the government had plans to transfer some of the Crop Insurance Fund into the Consolidated Fund, Shirley attempted to explain the theory of 'comparative advantage'. Pushing the line that, as a country, we should not try to do everything but should stick to what we do well, Shirley maintained that the protection tobacco growers were seeking amounted to a government subsidy. Unconvinced, growers pressed Shirley for answers on what alternative crops he could suggest for areas such as Dovedale and the Motueka Valley. They were not impressed with his response that he would not presume to tell farmers what they should grow.[37] Dissatisfied growers had few options but to get on with the new season and hope for the best.

Despite rumours of the government's designs on the money, the federation received the balance of the Crop Insurance Fund early in 1987 and began seeking advice on the most profitable investment for the fund. After recommendations from financial advisers such as Barclays and Fay Richwhite to invest in their unit trusts and commercial real estate, Ernie Drummond opted to hold the fund in Treasury bills. This decision paid more than a financial dividend – it saved the fund from complete disaster in the sharemarket crash of October 1987. When Treasury bills were redeemed in March 1989 the fund was valued at around $1,500,000. It was estimated this would have been slashed to no more than $300,000 if the advice of some investment experts had been taken.[38] Compensation for crop damage was slowly raised from its minimal level to over 50 per cent of the district average price. In time, the fund not only provided compensation for climatic damage but underpinned the cooperative's ability to offer growers the benefits of bulk purchasing.

In the 1987-88 season grower numbers were down to 100 producing from 544 hectares and the domestic order was now just over one million kilograms. With the season severely affected by wet weather during harvesting, the crop was light and many growers failed to reach their contracted production. Ironically, the shortfall was made up from in-store over-quota leaf from the previous season, a practice long desired by growers. Although acreage was now a fraction of its previous size there were still seasonal problems in acquiring sufficient satisfactory workers to harvest the crop. As the number of workers needed had fallen in recent years many growers began to let out their workers' baches on a permanent basis. Many were now unable to offer their workers accommodation. Local newspapers reported workers living in makeshift camps along river banks: regular police visits to one of these camps saw it christened Gordon's Camp after the local sergeant, Gordon Moore. Living in tents, polythene shelters and cars, the would-be workers survived by pooling their resources. Some were in difficulty as they had come to the district too early for both the fruit and tobacco seasons. The situation was far from satisfactory.

About the same time the phenomenon of the 'house bus' became a feature of the

Long-time industry supporter, W.E. (Bill) Rowling, M.P. for Buller, addresses the Federation's AGM 1974.

The 'Fairways' Hi-Trac tractor, especially suitable for cultivating tobacco.

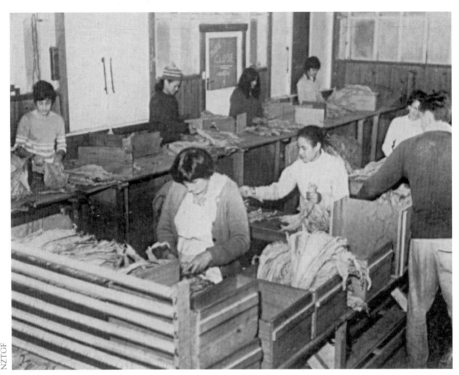
Hand grading remains into the 1970s although loose-leaf buying will soon be introduced.

Irrigation plants are now commonplace and have affected production markedly. These early units required the pipes to be moved every few hours, another arduous task.

Numbers of Fijian men travelled to New Zealand to work in the tobacco harvest between 1971 and 1981. Growers found these workers generally reliable and more likely to remain for the whole season.

R.R. (Roy) Rance, secretary of the Federation through the crucial years of the change from contracts to quotas and the restructuring of the early 1980s.

Company involvement in the local sporting and cultural scene was widespread. Peter Wild, Motueka manager of Wills, presents the Benson and Hedges trophy to the winning golfers.

Family labour is still a vital ingredient on many farms. Bulking tobacco on a Motueka Valley farm.

Fijian workers were appreciated for more than just their work ethic.

New kiln complexes of the 1970s featured the lower down-draught kilns.

Not all natural disasters were covered by the Crop Insurance Fund. This modern complex was flattened by wind.

Sticks were replaced at last. Green leaf is now held firmly in clamps which hold the equivalent of about ten sticks.

Technology marches on but smokos don't change much.

No more pipes to move! The travelling irrigator arrives.

Looking down the Riwaka Valley towards Motueka. Tobacco covers most available land c1975.

TOBACCO GROWERS CABARET EVENING

The time has come the walrus said, to think of many things— of shoes, a dress, a swish hair-do, baby sitters and rings — on the telephone to friends — who grow tobacco — to arrange an evening on May 25th to congregate with fellow growers.

A preliminary notice only and further details thru' press and Journal.

The Federation's business was not all serious. Annual cabarets were held for several years during the 1970s, each Ward taking its turn at the organisation.

J.C. (John) Hurley, Tobacco Board member 1965–1981, Federation president 1976–1981.

G.H. (Gerald) Hunt, Tobacco Board member 1966–1986, Federation president 1981–1987 and 1990–1995.

John Hurley, Bill Rowling and Gerald Hunt confer outside a Federation meeting.

Tobacco farmers take to the streets of Motueka to protest the government's plans to restructure the industry 1981.

Federation president, John Hurley, displays a graphic picture of 'who gets what' from the sale of tobacco products.

Some of the hundreds of growers and supporters who gathered at Memorial Hall to hear industry speakers discuss the government's proposals.

The final load of tobacco processed in Wills' Motueka plant leaving for further processing at the company's Petone factory.

W.D. and H.O. Wills' bond store being dismantled for removal by its new owners.

E.S. (Eric) Highet, Tobacco Board chairman 1978–1985. Highet steered the Board through one of the most difficult periods in its history.

E.W. (Ernie) Drummond, Federation president 1987–1990 and principal founder of the N.Z. Tobacco Growers' Cooperative Society.

Researchers and industry representatives gather to mark the end of tobacco research at the Riwaka Research station

Sowing seeds in the traditional way for the final season 1994–95.

The clocktower as remembered by locals and visitors alike.

Hand-picking for old times' sake – final season.

The last kiln ready for drying at Fry's farm, Riwaka.

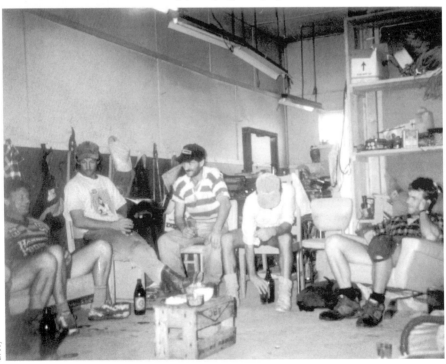

Harvest over for the last time.

Trucks lined up at the buying shed, July 1995.

The last bale is unloaded for inspection.

Motueka in the 1990s – a town built on tobacco?

Abandoned kiln in the Dovedale Valley.

harvesting season. With it came the problem of where such workers and their vehicles should be accommodated. Local authorities battled with the difficulty of a type of traveller whom everyone agreed should live somewhere but preferably not near them. Recreation reserves and country domains were suggested, as were existing camping grounds, but the administrators were not keen to cooperate to offer the required controlled living area. The local perception was that house-bus people were undesirable, had unsanitary living habits, owned ferocious dogs, and were inclined to petty crime. Some pip-fruit growers were quick to see the advantages of offering the limited facilities needed by house-bus and house-truck owners, but tobacco growers, on the whole, were less adventurous. The problem of workers and their accommodation remained.

Minor grumbling was also heard over the imposition of 'user pays' principles within the DSIR. Growers still contributed a significant amount towards tobacco research through levies but they were beginning to have doubts on the priorities adopted by the research station, and on the cost-effectiveness of what was being done. The Riwaka grower and Tobacco Board member Jack Inglis believed that heavy overhead costs were being loaded onto the research budget, thus reducing the actual spending on practical research. It was also noted that the industry was being charged for the use of facilities it had initially provided. From 1987 the DSIR required the industry to work towards total funding of specific tobacco research, agreed at $50,000. In 1987 the industry's share of this sum amounted to $23,755, rising to $25,818 the following year then progressing to 100 per cent of an agreed programme from 1989. To some this was an unfair situation as no other horticultural crop was required to totally fund its own research.[39]

Within these financial limitations the time spent on tobacco research was less than the equivalent of one scientist per year. Improved varieties remained at the heart of the research programme, with CRD Riwaka 88 being released in time for the 1988-89 season. This variety, like others specifically designed for local conditions, raised yields to unprecedented levels – there were reports of growers eventually producing well in excess of 3000 kilograms per hectare by the 1990s. In addition, the research station produced the necessary seed for the local crop. In 1988 this was valued at $2000 per kilogram.[40] From its beginnings as a facility set up mainly to service the tobacco industry the research station now focused on a wide variety of crops. Tobacco was rapidly becoming a minor part of the work of the station.

As varieties and yield had changed dramatically over the past twenty years, so had cultivation and harvesting practices on the farm. While planting techniques had remained consistent, laterals were now chemically controlled and 'topping' (removing the flowerheads) was carried out by rotating cutting bars mounted on tractors. Irrigation had progressed from the frequent manual shifting of pipes to underground mains servicing mobile rainers. The system used water more efficiently, could be working within thirty minutes, and required only one person to operate. Picking was almost entirely mechanical and bulk curing was the major method of drying tobacco. Grading of the cured leaf was no longer required and leaf was submitted for sale loosely sorted. With companies having relaxed their grading standards, a bulk kiln of approximately 100 clamps could be graded and crated for sale within three hours – previously this task would have taken up to two days, using six or seven workers. Over a period of twenty-five years the labour required to harvest and load a kiln of tobacco had dropped from twenty-plus to five or six. For many farmers this meant they could source much of their labour locally with only a minimal number of casual, outside workers. This was a significant development after decades of struggling for sufficient labour. In addition, it was clear that individual growers had cut their costs to the maximum degree. Any further reductions would need to be on an industry-wide basis.

By late 1987 the executive had established a growers' cooperative, registered as an industrial and provident society. This arrangement would enable the cooperative to make bulk-purchasing arrangements and to distribute any financial benefits amongst its members. The first meeting of the New Zealand Tobacco Growers' Cooperative Society Ltd was held in January 1988, with plans being presented to deal with some of the current problems in the industry. But in April 1988 further shocks rippled through the district. W.D. & W.O. Wills warned growers that, from the following season, it would pay only the ruling world price for domestic tobacco. At the time this was NZ$4.60 per kilogram, a price completely uneconomic for New Zealand growers. Wills claimed that the current price for domestic tobacco involved a premium of up to $2 per kilo, or $30,000 per grower. Rothmans, while willing to pay this premium in the short term, stated that it was paying a premium of $60,000 per grower. This would continue only while the domestic crop retained the advantage of stability and consistency of supply.

Ian Ross, managing director of Wills, spelt out his company's position and said it was up to growers to decide if it was worth growing tobacco.[41] Tom Inglis thought growers might as well end the industry now as linger on any longer. If the government could be persuaded to pay out $10,000 per hectare in compensation growers could get on with alternative crops and forget the discouraging results in tobacco. While many might have agreed, the cooperative decided to press onwards. Ernie Drummond presented a comprehensive plan involving a joint committee of growers and manufacturers to work on innovative options for the future. These options included joint processing of the crop in Motueka, further cost savings on the farm and the development of new crops.

Despite these joint discussions, Wills maintained its stance on the price of domestic tobacco. The company would need to know by 1 June 1989 whether growers would meet their conditions in time for the 1989-90 season. At a special meeting of growers in July 1988 growers expressed general dissatisfaction with the whole industry. There appeared to be no viable alternative cash crops in sight and many were struggling with the concept of a completely deregulated situation. At the September AGM confusion reigned amongst growers as their executive sought direction. The cooperative had been cautious in its first year and some opportunities for savings had been missed. A number of members saw this as indecision but it was obvious that most were unsure of exactly what they did want. The huge uncertainty of alternative crops haunted growers, who had known nothing but a crop which, despite constant setbacks and battles, provided a relatively secure and predictable return. Tobacco growers had never needed to do market research on their product, or to conduct feasibility studies. It was hardly surprising, therefore, to hear one prominent grower suggest that a professional adviser be engaged to decide what alternative crops growers should opt for, what the prospective market would be and then tell growers to get on with it!

The uncertainty was understandable. The realities of leaving the industry were by now obvious to remaining growers. Diversification into other crops showed variable results: kiwifruit was now experiencing its own traumatic downturn, newly planted pip-fruit was yet to begin producing and niche crops such as herbs and essential oil plantings were still to be proven. Few, if any, had moved into growing flowers. Family farms which had been devoted to tobacco for up to three generations were now likely to be providing subsistence living. Many rural women were working off-farm to supplement the household income or even to subsidise farm operations. Some of those who had accepted the special assistance grants found it difficult to cover the capital costs of establishing new crops; if the new crop failed to deliver satisfactory returns, the position quickly worsened. In many cases the size of tobacco farms made them uneconomic for other uses. A number were sold as hobby units, thus removing the prospect of more economic use of the land, a prime reason for restructuring in the first place. A few growers had taken advantage of the buy-back provisions and re-entered the industry, but the growing sector was never again to involve more than a fraction of its former strength. To remaining growers it appeared their position was weak, whether they stayed in or got out.

Meanwhile, the cooperative was shaken by the discovery in 1989 that, although the new growers' organisation had been duly formed and constituted, the old federation was still in existence. After a request from David Caygill, the federation had been re-registered as an incorporated society and, failing any moves to de-register the organisation, it remained a legal entity. Unaware of this, the executive believed, in any case, that the new cooperative would automatically replace the federation in all respects. The cooperative executive had therefore not conducted annual meetings of the federation nor prepared and submitted annual accounts. It soon transpired that legally the assets of the federation, which had been 'transferred' to the cooperative, remained the property of the federation. The cooperative had 'been trading and investing funds ... that it apparently does not own'.[42] Confusion reigned as lawyers struggled to sort out the exact position and what should be done to straighten things out. Generally, the best advice the embattled executive received was to continue with both organisations, using the cooperative as a trading vehicle and leaving ownership of assets vested in the federation. The question of membership complicated the matter and was to explode into a major confrontation in the early 1990s.

Through the late 1980s grower numbers remained steady at around 120 but early in 1989 the ground began to shake. Wills firmly restated its decision to offer prices in line with international tobacco markets. Although world prices had now risen to $5.50 per kilogram this was still well below an economic level for local growers, and when Wills offered a maximum of $5.60 for the crop of 1989-90 its growers were unable to accept. Rothmans' price was also marginal at $5.95. Growers could not make a living at these prices. Frantic negotiations took place in an effort to retain Wills' presence, including joint processing with Rothmans, but a mutually satisfactory price could not be agreed. The cooperative suggested leasing the Rothmans plant and taking over all processing itself. Again, agreement could not be reached; growers were divided and uncertain.[43]

Wills initially supported experiments in 'hand-stripping', which produced lamina and stems in the form it required, indicating that the company would be prepared to purchase 150,000 kilograms of this product. With the increasing popularity of roll-your-own tobacco, and the particular suitability of New Zealand leaf for this use, if local growers could supply hand-stripped leaf at a competitive price there would be definite hope for Wills growers. The experiments were carried out on the Motueka Valley farm of Len Bolger, but even though the results were successful the company withdrew its interest. Company personnel in Motueka gained the impression that the hand-stripping option had been offered on the spur of the moment by Petone-based officers and that over-riding instructions to close down local operations were coming from Wills' London head office. This was especially difficult for long-time company officers like Bert Black who found his position compromised. He had been completely unaware of the company's plans to withdraw from the local industry until February 1989, when rumours from the Petone operation began to circulate. Now Black felt growers would judge him guilty of encouraging them to continue growing when his company was so close to pulling out. Bert Black's forty years of service to Wills ended in disappointment and embarrassment and he was left feeling almost as bitter as his growers.

The cooperative desperately sought solutions to the imminent disaster. Approaches were planned to Phillip Morris, now based in Australia, to ascertain whether this company would be willing to purchase New Zealand leaf. The question of minimum domestic usage was put, somewhat forlornly, to Ken Shirley and his colleagues, along with proposals for suspensory government loans for diversification. In a paper to Shirley, Tom Inglis set out the possible costs of diversification. If the government agreed to finance the transition, Inglis believed it would come out 'smelling of roses'.[44] Hopes for such loans foundered on the government's knowledge that a number of those who had re-entered the industry under the buy-back provisions had not repaid their assistance grants. It was alleged that, while some had repaid all or part of their grants, others had not repaid a cent.[45] The Minister of Commerce, David Butcher, was clearly unimpressed with this when responding to a request for further assistance in 1989. Butcher reminded local MP Ken Shirley that growers had made commercial decisions to either remain in or re-enter the tobacco industry and must now reassess their positions.[46]

There was to be no last-minute reprieve for Wills growers. In July 1989 it was finally clear that the company would not be buying New Zealand leaf grown in the coming season. The Motueka plant would be closed, machinery dismantled and the buildings sold. Words could scarcely express the disbelief, both amongst growers and in the community, as the company which had been at the centre of tobacco growing since 1926 left the district almost overnight. Granny Wills had become Dame Misfortune. She was leaving, and leaving in circumstances which were to fuel a family feud with her growers.

NOTES

1. *Nelson Evening Mail* clipping, c. July 1980, Tobacco Growers' Federation files.
2. Broad, H. 'Sold Down Tobacco Road', *Straight Furrow*, 22.8.80, p.3. When the study group's report was made available, in 1984, the Tobacco Growers' Federation found it contained many recommendations with which it agreed.
3. *Nelson Evening Mail*, 15.7.80.
4. *Weekend Star*, 19.7.80.
5. *Nelson Evening Mail*, 19.7.80.
6. Ibid, 29.7.80.
7. Ibid, 15.7.80.
8. Ibid, 18.7.80.
9. Broad, op cit.
10. *Nelson Evening Mail*, 24.7.80.
11. Ibid, 31.7.80.
12. Ibid, 25.9.80.
13. Ibid, 25.9.80.
14. Ibid.
15. Tobacco Growers' Federation minutes, 20.8.80.
16. Parliamentary Debates, 1980, p.4314.
17. Letter from Doolan to Adams-Schneider, 6.11.80, Tobacco Growers' Federation files.
18. Tobacco Growers' Federation minutes, 17.12.80.
19. Notes of a meeting between industry delegates, the Minister and officials, 5.12.80, Tobacco Growers' Federation files.
20. Tobacco Growers' Federation minutes, 22.12.80.
21. Letter from Doolan to Highet, 31.12.80, Tobacco Growers' Federation files.
22. Tobacco Growers' Federation minutes, 11.2.81.
23. *Nelson Evening Mail* clipping, c. 1981, Tobacco Growers' Federation files.
24. *Nelson Evening Mail* clipping, c. April 1981, Tobacco Growers' Federation files.
25. *Evening Post*, 30.5.81.
26. Letter from Hurley to Adams-Schneider, 10.9.81, Tobacco Growers' Federation files.
27. Report of the Inter-departmental Study Group, 1980, p.39.
28. Tobacco Growers' Federation minutes, 30.9.81.
29. Tobacco Board report, 1982. In the event, manufacturers purchased all tobacco produced in the 1981-82 season.
30. Notes by A.L. Black, Motueka manager of Wills, c. 1990.
31. Tobacco Growers' Federation minutes, 29.7.83.
32. Notes by A.L. Black.
33. Tobacco Growers' Federation minutes, 29.9.83.
34. Ibid, 16.1.85.
35. Ibid, 30.9.85.
36. Tobacco Growers' Federation Annual Report, 1987, p.8.
37. Tobacco Growers' Federation minutes, 18.9.86.
38. Annual Report, NZ Tobacco Growers' Cooperative Society Ltd, 1988, pp.4-5.
39. Tillson, G., addendum (1995) to Notes on the History of Tobacco in NZ, 1980, p.7.
40. Federation and Cooperative annual reports 1987-90.
41. Tobacco Growers' Federation minutes, 13.4.88.
42. NZ Tobacco Growers' Cooperative Society Ltd minutes, 28.6.89.
43. Interview with former grower, federation and cooperative chairman E.W. (Ernie) Drummond, Dehra Doon, Riwaka, 2.12.96.
44. Paper from Inglis to Ken Shirley, c. 1989, Tobacco Growers' Federation files.
45. Tillson, G., addendum (1995) to Notes on the History of Tobacco in NZ, 1980, p.2.
46. Letter from Butcher to Shirley, 12.6.89, Tobacco Growers' Federation files.

CHAPTER EIGHTEEN

Death of the Dream

The departure of Wills swept all other concerns before it as for several months the cooperative attempted to deal with the aftermath. In August 1989 a meeting of Wills growers was held to discuss legal action against the company. In September Auckland lawyer C. A. (Chris) Dickie visited Motueka for discussions on possible legal moves. Dickie, from the firm of McVeagh Fleming, was associated with Auckland Queen's Counsel A. (Tony) Molloy. After viewing a televised interview with Ernie Drummond the pair had determined to assist growers in investigating a class action against Wills. The first stage, as proposed by Dickie, involved gathering and analysing information to allow Molloy to advise on the possibility of a successful claim.[1] The day after meeting with Dickie the cooperative resolved to set aside $120,000 for a fighting fund to contest the case. The cooperative assured growers that the costs of legal proceedings against Wills would be minimal, or even nil, and that any such costs would be met by the cooperative. Growers would not be asked to contribute any further funds.[2] Even so, some were dubious, holding an underlying suspicion of 'big-city lawyers'.[3]

Dickie and Molloy began collecting information to establish the basis of the growers' case for compensation. The initial aim was to plead that Wills had acted unfairly and inequitably in its withdrawal, without due notice, from the purchase of New Zealand tobacco. Growers might therefore be entitled to a payment in the nature of redundancy. Twenty-four Wills growers agreed to participate in the action which would attempt to prove that growers had been encouraged, even induced, by the company to continue growing tobacco in the early 1980s. It was alleged that new buildings and machinery had been purchased on the basis of statements and actions of company personnel during that time. Many growers had also re-purchased quotas after leaving the industry through the assistance grants scheme. Some claimed this buy-back was actively encouraged by Wills. Molloy now believed growers had a case for breach of contract, given the history of the relationship between Wills and its growers and the specific actions of the company over the past decade. Statements of claim were prepared, redrafted and redrafted again. The first year of legal activity went by but barely a shot had been fired: in fact there was hardly a sign of battle, let alone of a victory.

But another fight *was* brewing, this time within growers' ranks. The Pandora's box of federation membership, partly opened by the mix-up when the cooperative was formed, was now wide open. Since 1938 growers had been members of the federation by virtue of holding a licence to grow tobacco. When licences to grow were abolished in 1986 membership was informally based on the holding of a quota. Since quotas were also abolished, this was a questionable basis and it was accepted that changes to the federation rules were needed. The formation of the cooperative apparently removed this necessity. The rules of the cooperative society required members to be 'actually engaged in the growing of tobacco' and to hold a manufacturer's order for not less than 1000 kilograms of tobacco leaf. When the controversy over the existence of two grower organisations arose, some growers became concerned over their status. It was unclear whether those who had ceased growing, or had taken time out, were still eligible for membership of the federation which was still a legal entity. It was also thought that ex-Wills growers might be entitled to a share of Rothmans' order.[4] These were major concerns, given the uncertainty now hanging over the entire industry. Rothmans' order would not support more growers, and if the growing industry collapsed completely there were considerable assets to be disposed of – immense wrangles were foreseen if membership was in dispute.

Attempts to regularise the position became even more complicated when the election of federation officers in January 1990 was claimed to be irregular. In addition, a number of growers contested the rule changes passed at this meeting. Among these rules was a provision to set 1 April 1986 as the cut-off date for eligibility for membership of the federation. This would exclude any grower who had ceased growing before that date. The rule changes also allowed for any member of the federation to be elected to the executive: there was no requirement for a particular ratio of growers to non-growers. The changes were not agreed unanimously. More than a quarter of those present opposed measures which they believed would allow non-growers to participate in the affairs of the cooperative, and to receive benefits from federation funds and assets. These included the Crop Insurance Fund which now totalled almost $1.5 million.

Under the 1986 legislation the Crop Insurance Fund had been transferred to the federation, along with any 'liabilities' associated with the fund. The federation executive had interpreted this to mean that the fund was legally entrusted to it for the purpose of continuing to operate a crop insurance scheme specifically for tobacco growers. With the formation of the cooperative, a proportion of the fund had been used to make bulk purchases of fertiliser and sprays, such advances being repaid once growers received payment for the crop. Proposals to admit non-growers to the federation raised concerns that the funds accumulated by tobacco growers would be used to benefit those who had not contributed to them. The protesting growers sought a special meeting to re-elect the executive and strongly opposed the changes to the current rules of the federation. The protesters claimed, in fact, that the vote in favour of the changes had been less than the required two-thirds majority. This was found to be true and the changes could not be upheld. The position of the chairman, Ernie Drummond, and executive member Russell Cederman was less clear. Both had certainly ceased growing, being Wills growers, but Drummond's current position on the executive entitled him to re-election under the current rules. The protesters were by no means convinced.

A meeting of the federation executive in February 1990 considered the thorny issue.

The members were advised by their solicitor, Mark Wheeler, that membership of the federation consisted of 'any member who has ever been a member of the federation who hasn't died or been struck off the roll'.[5] Added to this was the long-standing difficulty of the rather informal determination of membership. With membership traditionally based on the holding of a licence to grow tobacco, no one had ever joined the federation, no subscriptions were ever paid, nor had anyone ever resigned. The executive was hardly cheered by this news, nor by the statement that the only way to have another election was for the present executive to resign.

What was needed, advised Wheeler, was a special meeting to adopt formal membership rules. The membership lists of 1979, drawn up as part of the basis for the special assistance grants scheme, could be used as a basis for determining who would be eligible to vote on rule changes. Discussion on the possible disposal of federation assets also brought warnings from both Wheeler and Gerald Hunt. Wheeler advised that strict guidelines were needed to ensure that no one could 'double-dip' in the event of the federation being wound up. Gerald Hunt firmly stated that any discussion of splitting up the funds should be avoided as 'it would send signals to all the membership who would want a piece of it'.[6] Hunt added that the funds of the federation should never be considered as property to be divided up, but should be used for the benefit of the people, both members and the wider community.

Growers protesting against the election and rule changes were invited to meet with executive but were advised that there would be no re-election of officers. The protesters explained to the executive that discussions with the registrar of incorporated societies had revealed that the federation membership may be irregular. The group, which included A.C. (Tony) Fry and B.Y. (Betty) Fry, requested the appointment of an independent arbitrator to resolve the issue. The chairman, Ernie Drummond, asserted that tobacco growers were not disadvantaged by enlarging the membership of the federation, but tartly added that they would certainly be disadvantaged if crop insurance claims could not be paid because of disputes and litigation brought on by the protesters. After discussion, the executive decided not to agree to the request for an arbitrator, largely on the grounds that a decision from such a person could not be binding and could as easily be challenged as the present situation. The matter was further complicated by a deputation, at the same meeting, of ex-growers seeking to remain members of the federation. Led by John Hurley, these growers did not want to see rule changes eliminate them from membership. In addition, Hurley sought assurance that the Crop Insurance Fund would not go to the last few growers. The federation solicitor, Mark Wheeler, intervened quickly to advise that 'any thought of the federation assets going to the last few growers ... should be stopped smartly'.[7] To the outside world, tobacco growers were valiantly struggling to survive as their industry crumbled, but internally the unseemly battle over the federation's entrails had already begun.

The executive sought high-level opinion from a Wellington barrister, Colin Carruthers. It specifically asked for advice on the status of the Crop Insurance Fund and the legality of the election of officers. Carruthers advised that the Crop Insurance Fund was free from any 'trust' implications and could be used by the federation for whatever purpose it chose. As to the election of officers, his opinion was that the 'unhappy' wording of the federation rules was at the heart of the problem. While awaiting Carruthers' opinion, the executive resolved that the only fair way to establish the true membership of the federation was to begin from the 1979 membership list. From this basis, it would be necessary to notify all members of a special meeting to tidy up the rules of the federation and to decide on the basis of continuing membership. Even this process was fraught with difficulties but it would serve to allay fears that non-growers were about to take over the affairs of the federation. As the heat of the battle began to diminish, wounded were left in its wake. Ernie Drummond resigned as chairman in May 1990 with health problems which were no doubt aggravated by the constant combat of the previous year. In the election for a successor, Gerald Hunt found himself once more at the helm of the industry, an industry in an even more critical state than when he had left in 1987.

By the end of 1990 the membership of the federation had been placed on a rational basis. There was little protest, although no doubt the changes did not satisfy all. Membership would now be by subscription and members could be removed from the list of members if subscriptions were not paid. Discussion on the disposal of federation funds in the event of its demise also faded away. Attention could now turn to the question of alternative crops for those forced out of the industry and to the continuing legal case against Wills.

Wills' withdrawal had left only fifty-six growers, producing from an area of 339.8 hectares. Production was now limited to the amount required by Rothmans, and this company gave firm assurances that growers would receive at least two years' notice of any move to cease buying domestic tobacco. Overproduction was still a concern as many growers continued to plant greater areas than required to produce their contracted amount of leaf.[8] Rothmans introduced financial penalties for overproduction in the late 1980s and these were strengthened in the 1990s. The company began to divide contracts on the basis of current production in one contract and carry-over leaf in another. The contract for production in the current season would be reduced by the amount of carry-over leaf held by the grower – another version of the reducing quotas employed in the late 1970s. Quantities required by Rothmans continued to decline as the 1990s progressed, making New Zealand leaf even less competitive. In addition, since Wills was no longer required to purchase domestic leaf, its products were now more competitive with Rothmans.[9] As the downward trend continued, growers intensified their efforts to diversify, with seminars being held on a wide range of alternatives. Various advisers addressed meetings on the pros and cons of herbs, essential oils, pip-fruit, grapes and flowers, especially calla lilies. Other suggestions included nashi pears, feijoas, citrus, nuts and organic vegetables.

The concept of grape-growing was not new. Successful vineyards were already producing quality wine in the Nelson district and in 1990 Tony Molloy had floated the idea of a wine cooperative as a solution for Wills growers. A wine producer himself, Molloy believed it might be possible to acquire an overseas joint-venture partner to fund the capital needs of the project. He had already discussed soil suitability with a Ministry of Agriculture and Fisheries consultant and the results appeared promising.

After a preliminary report, the matter would be discussed with an international wine scientist.[10] Neither the federation nor the cooperative appear to have responded to the idea – the matter was not discussed at executive meetings.

Tobacco growers were still, on the whole, small holders and were unfamiliar with the concept of growing crops under a cooperative system. Many had tried and failed to organise them to produce and market their tobacco in this way. Despite the best intentions of Ernie Drummond, the Tobacco Growers' Cooperative Society was a cooperative in name only, as growers did not produce and market their tobacco on a cooperative basis. The cooperative was largely a means of using the Crop Insurance Fund to obtain financial savings for its members through bulk purchasing. It did not, as Drummond had hoped, initiate diversification on an industry-wide, entrepreneurial basis. In the event, growers leaving the industry were forced to make individual decisions on the future direction of their land use – regardless of the degree of urgency, a number still found the decision to enter new ventures difficult. Such decisions involved taking commercial risks with land which had been in families for several generations. Failure would mean more than financial loss to these farmers.

Much of the controversy surrounding the federation had now dissipated. By late 1991 the cooperative was used only for trading purposes and not all growers supported the bulk-purchase arrangements obtained. As well, the cooperative had avoided some bulk arrangements in order to retain the support of local merchants. The 1991 AGM of the federation attempted to inject an air of optimism. The chairman, Gerald Hunt, reported that the industry's outlook was positive, with the level of the New Zealand dollar and increased world demand for tobacco restoring some competitiveness to the local product. The industry now offered employment to approximately 120 permanent workers and 400 seasonals. Of increasing concern, however, was the impending ratification of the CER agreement, which would see Australian-made products competing on the New Zealand market, some manufactured from cheap imports. Legislation banning advertising of tobacco products in New Zealand publications would also advantage Australian companies, which could promote their products in magazines and periodicals freely available in New Zealand. Ending on a positive note Hunt suggested that when economic conditions finally improved, the good relationships now existing between growers and their buyer would ensure a beneficial future for the industry.

With little benefit now being obtained from the existence of the cooperative the organisation was dissolved in July 1992. No fanfare accompanied this move – the tension and controversy of only two years before had apparently evaporated. The ongoing concern was now the length of time the Wills case was taking to reach a court hearing.

The judicial process was grinding 'exceeding slow'. Wills had appointed R.T. (Russell) Feist, of the Wellington firm of Tripe Matthews and Feist, to represent it and had secured the services of George Barton QC. The growers' own QC, Molloy, was happy with the appointment of Barton; he felt that both parties would cooperate over presenting documents and other relevant information. There should be no unnecessary delays ahead.[11] But the growers' own case was delayed by conflicting information from growers, which caused Dickie and Molloy to advocate significant changes to the thrust of the case. Much of the evidence on which the growers based their case was undocumented and involved statements allegedly made by company personnel. Apart from the history of annual contracts between Wills and its growers there was little solid evidence that the company had acted unfairly in its withdrawal. The claim for compensation in the nature of redundancy could be difficult to prove. Dickie now advised that a case for breach of contract was a safer course and would be easier to argue.[12] In the interim, Russell Feist had advised that Wills intended to actively defend the case.

Soon after preparation of the case began Wills' long-time Motueka field officer, leaf buyer and local manager, Bert Black, indicated that he would assist growers by providing information. Black outlined the way in which he himself had been led to believe that the company would retain its involvement with New Zealand growers. He had been present at the meetings of the early 1980s when successive general managers of Wills had given the impression that the company was here for the 'long haul'. Black had been requested, in 1983, to delay his retirement beyond the official retirement date, as the company wished to appoint and train a suitable successor. This trainee, A.J. (Tony) McLean, was told in 1986 that he would be in Motueka for about ten years, the inference being that Wills would continue to purchase local leaf for at least that period. Black had been stunned to be told, early in 1989, that the initial price offered to growers for the coming season should be disregarded. The price to be offered would be more than a dollar per kilogram lower. To Bert Black this appeared to be a deliberate attempt to set a price at which New Zealand growers would be forced out of production.[13]

Bert Black's evidence was regarded as crucial as the case against Wills developed. But the case was still developing only slowly. Wills presented its Statement of Defence in August 1992. In essence, the company agreed only with the obvious facts of its existence and the existence of certain growers, but denied, or claimed no knowledge of, all other points made in the growers' case.[14] Wills then served a 'notice for discovery', which required the filing of all relevant supporting documents with the court. Growers now needed to provide documented evidence of every activity involving themselves and the company. Molloy was aware of the escalating costs of the case and assured the federation that he did not intend to add to the growers' burden by presenting frequent accounts.[15] The costs were indeed escalating. In December 1992 the federation had already paid $82,800 in legal expenses and it was estimated that the cost to bring the case to court could be a further $85,000 plus GST.[16] The complexities of the case were also outlined at this time and it was clear that growers had undertaken a mammoth battle.

Long before this, however, the growers involved had already become frustrated by the delays and apparent lack of action. In August 1992 the 'Wills group' demanded that the federation address the situation immediately. 'After three years and a small fortune in costs'[17] the group contended that some growers were still in limbo, unable to make critical business decisions as they awaited the outcome of the case. There was little cheer for them in the news that a court hearing was unlikely before the second half of 1993. The hearing was no nearer in late 1993, but the costs continued. By October 1993 a total of $134,345 in legal fees had been paid – it was now estimated that a further $100,000 was needed to bring the case to a conclusion. The federation secretary wryly pointed out that estimates had proved to be less than half of the actual amount paid.[18]

In addition to their their frustration over the three years of legal meandering, some Wills growers were still coming to terms with the withdrawal of 'their' company. Speaking to a federation executive meeting in May 1993, Graham Emerre spelt out what Wills had meant to growers. The company had built up a family image in its dealings – 'growers *never* felt Wills would leave'.[19] Most had thought it more likely that Rothmans would be the first to pull out. Wills had been a company of substance, solid and stable, with a history to back it. Growers had been loyal to the company and were devastated by its departure.

Growers' feelings notwithstanding, the legal process dawdled on. In September 1993 growers were informed that a hearing might now be set for early in 1994; in October 1993 they were asked to approve a further $10,000 towards legal costs. No hearing eventuated either early or late in 1994. Midway through the year, Molloy floated the idea of a negotiated settlement with Wills, revisiting the idea that the Motueka growers could convert their tobacco farms to vineyards. While continuing the necessary formalities of the case against Wills, Molloy prepared an extensive proposal for presentation to the tobacco company. Written under the name of Gerald Hunt, the proposal offered Wills the opportunity to enter a joint venture, with a possible overseas backer for the growers, to establish grape-growing in the Motueka area. It was suggested that Wills might fulfil its obligation to its former growers by providing a building suitable for a winery.[20]

Predictably, Wills rejected the proposal. George Barton had now been replaced. The new QC was said to be intent on dragging out the litigation as long as possible – the multinational could afford to prolong the case, the growers could not. In April 1995 the federation's new secretary, Darien Beckett, requested information on where the case stood, plus a further financial forecast. The reply from McVeagh Fleming brought consternation and dismay. The cost to bring the case to a completed hearing was now estimated at $317,000.[21] Hard on the heels of this bombshell came the statement that the case was breaking new ground and that success could not be guaranteed.[22] It was plain that the federation could not commit any further funds to the case – its ability to do this at all had already been questioned by some. The Wills growers were doubly devastated. They had lost their traditional source of income when the company left town; now they stood to lose any possible benefit from five years of case preparation.

As it turned out, devastation would not be limited to Wills growers. In August 1993 Rothmans announced that it would review the feasibility of domestic tobacco on an annual basis from the 1994-95 season. This did not amount to a definite withdrawal date but few doubted the probable outcome. Unless local leaf could compete with imported tobacco, it was obvious to remaining growers that their industry's days were numbered. Though the company had honoured its promise to give at least two years' notice of any possible withdrawal from the local market, the announcement was received with dismay. Diversification had taken place on a large scale in recent years, but there remained a significant amount of land for which no alternative use had yet been found. Similarly, it was unlikely that other uses would be found for the buildings and plant peculiar to tobacco cultivation and curing. Certainly no other crop would provide the cash injection the local community had enjoyed so long from the tobacco industry.

Some commentators took an optimistic view. Tasman District mayor Kerry Marshall was hopeful that conditions might change to allow the local crop to compete. Growers had been nervous for several years, he stated, but the way was now clear for growers to make decisions on their future.[23] Gerald Hunt also believed the news did not necessarily mean Rothmans would be pulling out of the district. World conditions could change over the following two years and the present low international prices might not be repeated. All the talk and positive thinking in the world, however, did not change the fact that the cost structure of the New Zealand growing industry could not be pruned to the level required to compete with low-cost producers. There was little doubt in the minds of most growers that 1994-95 would be the last season of tobacco growing for them. This was confirmed in August 1994 when Rothmans announced that it would not purchase domestic leaf beyond the 1994-95 season. The long and often dramatic story of commercial tobacco growing in New Zealand would at last come to an end.

Knowing it was their final season, growers found it difficult to maintain the level of commitment as they went about preparing seed-beds and cultivating ground for the last time. The remaining fifty-one growers planted out 189 hectares to service a company order of 600,000 kilograms and achieved actual production of 530,219 kilograms. Much media attention was concentrated on the last harvest, with regular visits from television crews and journalists. Besides the brief national television coverage, lengthy articles on the history of the crop and the probable outlook for the district appeared in magazines and newspapers around the country. Growers were finally getting the chance to tell their story to the nation, but too late to have any saving effect. The death of an industry producing what had become an unpopular product received little real sympathy from the public beyond a passing murmur as the images of disgruntled and depressed farmers flitted across their screens. Many seemed of the opinion that other more desirable crops would be a far better use of the land. Tobacco farmers knew that although they might plant their fields with alternatives, no crop existed which would fill the looming holes in their annual budgets.

But the harvest proceeded, albeit to unaccustomed attention, and television crews were again on hand to record the final buying. The usual motley assortment of trucks, large and small, multi-coloured and multi-aged, lined up day after day with their golden loads. For most of the buying period the procedure was as it had been for decades. Trucks came and went, bales were weighed and classified, prices allocated and the leaf began its journey through the mixing and redrying machines of the busy factory. But the final day of buying, Tuesday, 27 June 1995, brought sober faces among both growers and company personnel. Cameras flashed and jokes cracked as the last bales were delivered but the air of finality was evident. Heads shook in disbelief as the realisation sank in that this was the last act of what had been a vital part in the lives of all concerned, whether involved on the farm or in the factory. 'I'm really feeling it now,' commented factory manager Geoff Tillson, who had begun his working life with Rothmans in 1961 – 'It's really over.'[24]

But like all good stories it was not really over at all. As Rothmans staff began the process of dismantling machinery and plant tobacco growers provided a post-script which would see the industry end in the controversy which had surrounded it from its infancy. Once the final buying was over attention turned to the business of winding up the federation and the ultimate disposal of assets. The Crop Insurance Fund was now valued at approximately $800,000, and the federation also owned its King Edward Street building. Growers began a series of meetings to consider their options. The community, aware of the implications, looked on with interest.

The executive held discussions with the federation solicitor, Mark Wheeler, and considered all options. These included donating the funds to a chosen cause or causes, splitting the funds between remaining growers, and setting up a business enterprise which would return dividends to shareholders. With the Rothmans plant now on the market there was some interest in purchasing this for community or commercial use, but the executive felt it was not in a position to enter discussions. At the same time members of the local community were interested in acquiring the buildings for use as a community facility, and to ensure that the character of the landmark corner was maintained. Public meetings resulted in a feasibility committee charged with the task of examining possible uses and sources of funding to enable the purchase to proceed. It was soon obvious that the community would be unable to raise sufficient money and the project was abandoned. The building was sold to a private buyer and attention turned again to the federation and its plans for the Crop Insurance Fund.

Grower meetings in late 1995 saw all the old divisions reappear. Understandably anxious over their own futures, many final-year growers were conscious of the public perception that they were indulging in an unfair 'divvy-up'. Former growers added their voices. Some who had been out of the industry for years expressed strong feelings that, as the fund had been accumulated over decades by growers, many of whom were long dead, the ethical course would to be to use the money for community benefit. Theatrically, some asserted that 'graves would open' in indignation if present growers shared the fund out amongst themselves.[25] John Hurley again raised his voice, supporting the view that the fund should not be handed out solely to the last few growers. Hurley believed an acceptable compromise would be to split the fund, paying out half to growers and vesting the remainder in a community trust for the benefit of the local community which had supported growers in many of their past battles. Other former growers, while not offering specific proposals, were equally scornful of any intention to divide the funds amongst remaining growers. While not desiring any direct benefit for themselves, some were embarrassed at the perception that those growers still in the industry were intent on sharing out the spoils amongst themselves. It was seen as a sad end to an industry which had provided Motueka with a degree of wealth and a character unique to the area.

Former Wills growers were particularly bitter over the decision to divide the assets of the federation amongst the remaining Rothmans growers. The fund had been built up by growers of both companies and Wills growers felt it was now being unfairly distributed. This was not their only cause for bitterness. Wills growers had met in August 1995 and resolved to withdraw their case against the company. In signing authorities to cease legal claims, many added sharply critical comments against their legal advisers. Several reminded McVeagh Fleming that Chris Dickie had initially approached the growers, seeking to represent them. One grower suggested that, since the whole idea had been Dickie's, any costs claimed by Wills should be referred back to the lawyer.[26] Although this reflects understandable bitterness, it is not completely fair to Dickie; growers could have rebuffed his advances or ceased their action at any time. Ultimately, all twenty-four plaintiffs agreed to withdraw their claims.

Having made the decision to cease legal action against Wills, these growers felt they had ended the matter. On 14 September 1995 Molloy advised that he had reached agreement with Wills' legal representative, and that no costs would be sought. The long, messy affair could now surely be put behind them. But further trauma was to come. Two weeks later the company reversed its position and once again claimed costs from growers. McVeagh Fleming advised that Wills had seen newspaper reports regarding the disposal of federation assets to growers and had withdrawn its offer to waive costs.[27] The company's change of heart was apparently based on the belief that the federation had funded the legal costs to date and it should, therefore, pay costs 'owed' to Wills before dividing up assets amongst growers. This, of course, was not possible. Most Wills growers had no claim on federation funds and the federation had resolved that it would no longer facilitate the legal dealings. To the Wills growers this was the final insult: while the federation was proposing to split its assets amongst the Rothmans growers, the Wills growers might be forced to use their own funds to finalise the case. After three weeks of intense anxiety McVeagh Fleming again advised that a settlement, without costs, had been reached. The bitterness between growers would be longer in dissipating.

At the final meeting of the federation, on 13 November 1995, the members confirmed the resolution that '[b]ecause the federation can no longer achieve the objects set out in its rules, it is hereby resolved that the federation be wound up voluntarily'.[28] Having considered all possible options for the disposal of federation assets, the meeting resolved to distribute the available assets amongst the remaining members of the federation. In December 1995 accountant Mark Brown, acting as liquidator of the federation, paid out the bulk of the Crop Insurance Fund on the basis agreed. By some standards the amounts received were hardly worth all the fuss, but to growers facing an uncertain future they provided a degree of compensation. As they prepared their forward budgets many could see years of deficit ahead. Bank managers would need to be accommodating for some time to come. Whatever alternative they chose, tobacco growers would not make anything like the earnings received from their golden 'weed'.

The federation building in King Edward Street was sold in October 1996. Federation records, minutes and correspondence, some dating back to the 1940s, were sorted and packed. Essential material was sent off to the National Archives and the rest put into storage. In May 1997 the door of the federation office was closed and locked for the last time on the New Zealand tobacco-growing industry.

NOTES

1. Tobacco Growers' Federation minutes, 19.9.89.
2. Notice to Wills growers from NZ Tobacco Growers' Cooperative Society, Tobacco Growers' Federation files, 13.9.90.
3. Interview with former grower Alan Moss, Little Sidney, Riwaka, 10.12.96.
4. Interview with Ernie Drummond, 2.12.96.
5. Tobacco Growers' Federation minutes, 7.2.90.
6. Ibid.
7. Ibid, 27.2.90.
8. It was frequently expressed to the author that this practice characterised the entire history of tobacco growing in New Zealand. The practice was so widespread that it affected the ability of the quota committee to operate. Grower members of the board knew all too well how growers managed to bend the rules. Interview with Ernie Drummond, 2.12.96.
9. Tillson, G., addendum (1995) to Notes on the History of Tobacco in NZ, 1980, p.5.
10. Letter from C.A. Dickie to Tobacco Growers' Federation, 20.6.90, federation files.
11. Ibid, 29.1.91.
12. Ibid, 19.2.91.
13. Draft of evidence of A.L. Black, Tobacco Growers' Federation files.
14. Copy of Wills Statement of Defence, 31.8.92, Tobacco Growers' Federation files.
15. Letter from A. Molloy QC to Tobacco Growers' Federation, 26.3.92, federation files.
16. Letter from J.L. Clark, McVeagh Fleming, to Tobacco Growers' Federation, 17.12.92, federation files.
17. Letter from Wills growers to Tobacco Growers' Federation, 9.8.92, federaton files.
18. Notes attached to Tobacco Growers' Federation minutes, 30.9.93.
19. Tobacco Growers' Federation minutes, 31.5.93.
20. Copy of a proposal prepared by A. Molloy for submission to Wills NZ Ltd, 27.2.95, Tobacco Growers' Federation files.
21. Letter from McVeagh Fleming to Tobacco Growers' Federation, 13.7.95, federation files.
22. Letter from McVeagh Fleming to Tobacco Growers' Federation, 20.7.95, federation files.
23. *Motueka-Golden Bay News*, August 1993.
24. Notes taken by author at the final buying session. For Tillson, the story was not quite over as he was to spend the next few weeks dismantling machinery and plant ready for shipment to another Rothmans facility.
25. Interview with Ernie Drummond, 2.12.96.
26. Moss, A.F., Authority to discontinue legal action against Wills,1995. Quoted with Moss's permission.
27. Letter from McVeagh Fleming to Tobacco Growers' Federation, 27.9.95, federation files.
28. Minutes of final Tobacco Growers' Federation meeting, 13.11.95.

Raking Through Ashes

And so the dream died. The crop which sustained hundreds of families and workers for almost seventy years was ploughed under for the last time. Alternative land use and budget deficits are a reality for former tobacco growers. One lone grower, like a latter-day Charles Lowe, began growing and processing his own crop, intending to market his product independently. But one grower does not an industry make. It is now time to look back over the history of tobacco growing in New Zealand and to attempt to make some sense of its beginnings, its chequered history and its eventual collapse.

Despite the early settlers' modest success at growing, it is clear that no commercial tobacco growing would have become established in New Zealand without the impetus provided by Gerhard Husheer. Until his venture proved its success following the First World War there was no interest from other tobacco manufacturers in purchasing domestic tobacco. Even when W.D. & W.O. Wills became interested in the possibility of flue-cured tobacco production there was no surety that the company would have a long-term involvement. The National Tobacco Company, on the other hand, quickly developed a paternal relationship with its growers. Personalities like Husheer and Cecil Nash took on a legendary aspect to growers, and the company leant heavily on its image of the protector as domestic industry.

As history shows, Wills did remain a presence in the local industry and, ironically, came to be seen as the family company. Wills developed its own form of paternalism and secured from its growers a high degree of loyalty. Having begun its involvement by offering finance for kilns and buildings, the company expanded this to the provision of seasonal finance, field advice and budget supervision. With so much of their day-to-day life supported and controlled by the company, it is not surprising that growers developed a dependency which extended over several generations and which was difficult to break free from.

Analysing the role of tobacco companies in the production of tobacco in New Zealand is problematic, in that the companies can scarcely be discussed in isolation from tobacco growers. From the paternal to the purely commercial, from cooperation to aggression, from support to outright abandonment, company actions towards their growers cover the entire range. Most growers enjoyed good relationships with at least the local representatives of their company, but there were instances where company personnel were not welcome visitors on certain farms. The peculiarity of the relationship between grower and company is probably only understandable in terms of the annual contract system and seasonal finance provisions. Companies acted somewhat in the role of guardian, adviser and mentor; coming from backgrounds of struggling small holdings, many farmers found this security blanket very seductive and addictive.

Paradoxically, the independent leanings and geographical isolation of many growers were also factors in the struggles of the industry. Personal differences and district parochialism brought divisions and suspicions within grower organisations. Such divisions often disadvantaged and diverted growers in their real purpose – to secure greater control over their industry. This, of course, conflicts with the allure of the security offered by company paternalism. The contradictions inherent within these two concepts were never resolved. To the very end of the industry, and even beyond for some, there was a feeling that 'the companies ought to have looked after us'.

From a company perspective, New Zealand growers had certain benefits and drawbacks. As long as the benefits outweighed the drawbacks both Wills and Rothmans were willing to continue to purchase domestic tobacco leaf. Security of supply and price, plus the ability to monitor production methods, justified the premium of a higher price for domestic leaf until well into the 1980s. Once tariffs were reduced to below 12 per cent, and immense amounts of cheap leaf became available from elsewhere, it was increasingly difficult to justify the extra cost of purchasing New Zealand tobacco. Wills was the first to submit to these pressures and it was only a matter of time before Rothmans followed.

The role of the government is more difficult to analyse. Whatever form the company-grower relationships took, government positions on tariffs and the protection of domestic industries were vital. Moves in the 1920s to support tobacco production for export came largely from the efforts of the Director of Horticulture, J.A. Campbell. Charles Lowe's enthusiasm and ability to motivate others also played a part in the government's support of the ill-fated export venture. The brief flirtation in the Pongakawa Valley has already been described as misguided: the then government was clutching at virtually anything to alleviate the horrendous unemployment problem.

The introduction of regulatory legislation in 1935 can probably be attributed to a combination of exasperation with the continuous problems of the industry and the persistence of Keith Holyoake. The continual stream of delegations and deputations of tobacco growers to Wellington obviously caused some disquiet amongst government officials and led to a wearing down of official resistance to growers' demands. Keith Holyoake was a vocal advocate for his electorate; his efforts did him no harm at all on the wider political scene. Locally, Holyoake lost favour and was not re-elected in the 1938 election. The industry retained some political leverage as long as the local member was also a member of government.[1] With the entire growing industry concentrated within one electorate, it was difficult to marshall sufficient pressure unless there was a direct line into Cabinet.

Even with the 1935 legislation in place, it is unlikely that the tobacco-growing industry would have enjoyed the success it did without the outbreak of World War Two. War-time conditions of shipping constraints and the necessity to conserve funds created ideal conditions for the growth of domestic production. Only the shortages of materials and labour limited the degree to which growers could expand their production. By the end of the war the industry was well established, consumers were accustomed to a significant percentage of locally grown tobacco, and manufacturing operations were geared to continuing the existing patterns.

Legislation enshrining minimum prices, a district average price and, later, a mandatory domestic usage percentage, was vital for the survival of the New Zealand industry. Without these provisions the reality which faced growers in the 1980s and 1990s would have been evident decades earlier. The cost-plus method of determining the price of leaf also allowed a false sense of profitability. As long as growers recovered their costs, and received a margin of personal reward above this, the industry (along with many others, of course) was operating on a false basis. As the concept of the need for ever higher prices became embedded in the psyche of the federation and its representatives, growers were in danger of losing touch with the value of their product.

The change of government in 1949 brought the first indications that government support would not necessarily continue. Tariff protection was removed in 1951, imports surged immediately and the domestic industry suffered as a result. Growers were not helped by divisions within their own ranks. No tangible support came from their old ally, Holyoake, who was now a prominent member of the National government which espoused the removal of tariffs for trade purposes. It was only when the Labour Party again took office, in 1957, that protection was reimposed and expansion again occurred.

But the real boost for domestic growers was the arrival of Rothmans on the domestic buying scene in the late 1950s. The ensuing competition for growers brought unprecedented growth in grower numbers and in production. Only a few realised that this was a false boom and that the day of reckoning was inevitable. Even so, many of those who entered the industry at this time were able to establish themselves and go on to enjoy a relatively profitable life, retiring just before the final difficulties which brought the industry to a close. The arrival of Rothmans, whatever its apparent benefits, signalled a change in the relationship between companies and growers. A more aggressive, often litigious manner of dealing developed and would not be softened until the 1980s.

The problems resulting from the boom of the early 1960s forced a re-evaluation of the entire industry. The lengthy, often frustrating, Commission of Inquiry brought no results until a Labour government again occupied the Treasury benches. Even then, the resulting legislation did not bring the expected benefits of stability and security. The industry's long-time supporter, Bill Rowling, could only express disappointment and a feeling of foreboding as international pressures began to impact on domestic growers. The combination of financial pressures on companies and changes in government philosophy brought the first hint that the growing industry might be tossed into the current of market forces.

The deregulation proposals of the early 1980s were heartily endorsed by the incoming Labour administration. The Minister of Trade and Industry, David Caygill, clearly illustrated this by his statement to federation president Gerald Hunt that the government was 'not in the business of insulating New Zealand industry against economic reality'.[2] Economic reality for tobacco growers was a shrinking order from manufacturers and a very narrow range of alternatives. No further government assistance was forthcoming throughout the rest of the life of the industry.

It is possible to view tobacco growing in New Zealand as something of an aberration.

The climate and soil types are considered by some to be marginal and in fact the growing area was confined to a relatively small geographic area. The predominance of small holdings contrasts with the vast plantations of other tobacco-growing countries, as does the use of seasonal, untrained labour. The contract system of buying, which prevailed until 1974, was virtually unknown anywhere else. Many believed this gave tobacco companies complete control of the growers; others that it provided security. Certainly it contributed to the instability and fluctuation in grower numbers and production.

However the industry is viewed, whether as an aberration or as an industry which had its place and time in New Zealand history, its demise can only threaten the economic viability of the Motueka area. To a great extent the economic downturn which hit other rural areas in the 1980s bypassed Motueka. Locals counted themselves lucky to have such a varied economy, underpinned by horticulture, fishing, forestry and tourism. The income from tobacco, estimated at $6 million even in the final season, constituted a significant part of that prosperity. Today there are signs that the community is beginning to feel the effects of the first year without tobacco income. The Motueka Valley has lost much of its former immaculately groomed look – kilns stand awkwardly in fields with nothing to do. There is an unprecedented number of empty shops in the retail sector of Motueka. The prosperous look and feel of the area, for so long obvious to visitors, is beginning to slip. The clock tower still stands on its corner, naked now without the distinctive trade name. Negotiations continue on its future, with the local community anxious to preserve the forty-five-year-old landmark.

Those involved in the cultivation and processing of tobacco, both farmers and workers, have lost not only their livelihoods. Many have lost the pattern of their lives, the annual round of tasks and personal interactions involved in the production of the crop. A number have been forced to forfeit family land. Those who have known no other occupation struggle to maintain their sense of identity. The district itself has lost a certain uniqueness, a sense of character born of being the only tobacco-growing area in the country. Tobacco may have become an unfashionable and unpopular product, but the passing of its growing will long be mourned by the people of the Motueka area.

Notes

[1] Probably the only instance in which this did not apply was during the 1980s, when Ken Shirley was MP for Tasman. Although giving growers a degree of support, Shirley was unable to exert any pressure on Cabinet to alter or defer the terms of deregulation.
[2] Letter from Caygill to Hunt, 29.5.86, Tobacco Growers' Federation files.

Officers of the Tobacco Growers' Federation 1938-95

Chairmen

F.A. (Fred) Hamilton	1938-50
W.C. (Fred) Wills	1950-62
K.J. (Kos) Newman	1962-65
W.J. (Wally) Eginton	1965-66
R.W.S. (Dick) Stevens	1966-76
J.C. (John) Hurley	1976-80
G.H. (Gerald) Hunt	1980-87
E.W. (Ernie) Drummond	1987-90
G.H. Hunt	1990-95

Secretaries

N.J. (Noel) Lewis	1938-53
J.J. (Jeff) Bradley	1953-61
K.C. (Ken) Collins	1961-64
N.C.L. Holloway	1964
K.C. Collins	1964-66
W.G. (Willis) Bond	1966-72
R.R. (Roy) Rance	1972-86
M.D. (Mel) Cox	1986-94
D.E.C. (Darien) Beckett	1994-95

Members of the Tobacco Board 1936-86

(Board terms were from August to July)

Chairmen

Louis J. Schmitt	1936-53	Sir Jack Harris	1974-76
R.B. Tennent	1953-58	S.F. Ashby	1976-78
H.L. Wise	1958-59 (acting chairman)		(died in office, June 1978)
H.L. Wise	1959-69 (May)	E.S. Highet	1978-85 (September)
P.B. Marshall	1969-70	D.J. Gasson	1985-86
J.F. Cummings	1970-74		

Growers' Representatives

J.F. Balck	1936-43	H.E. Holyoake	1953-61
G.W. Relat	1936-39	S.E. O'Hara	1953-59
N. Rowling	1936-48		(died in office, June 1959)
H.A. Thorn	1936-41	M.A. Cederman	1959-61
F.A. Hamilton	1939-53	S.J. Emerre	1961-64
B.T. Rowling	1943-45		(died in office, June 1964)
M.H. Thorn	1940-50	J.R. Talbot	1961-71
R.W.S. Stevens	1945-51 and 1964-79	W.J. Eginton	1962-66
K.J. Newman	1948-65	J.C. Hurley	1965-81
	(died in office, May 1965)	G.H. Hunt	1966-86
W.C. Wills	1950-62	D.O. Cederman	1971-75
J.R.D. Drummond	1951-53	J.L. Inglis	1979-86

Manufacturers' representatives

R.B. Smith	1936-37 (resigned April 1937) W.D. & W.O. Wills
G. Husheer	1936-37 (resigned April 1937) National Tobacco Co.
K.A. Snedden	1936-37 (resigned June 1937) Consolidated Tobacco
A.H. Spratt	1936-41 Godfrey Phillips
E.M. Hunt	1936-47 (appointed April 1937) W.D. & W.O. Wills
C.C. Nash	1937-44 (appointed April 1937) National Tobacco Co.
A.F. Bell	1937-41 Nelson Tobacco Co.
W.R. Olliver	1942-48 Nelson Tobacco Co.
C.M. Paynter	1941-48 Godfrey Phillips
T.P. Husheer	1944-53 and 1954-58 National Tobacco Co.
F.A.L. Hunt	1947-59 W.D. & W.O. Wills
F.E.J. Jeffcott	1948 (died in air accident, March 1948) Godfrey Phillips
J.G. Lisman	1948-53 Company uncertain
T.F. Varley	1949-53 Godfrey Phillips
F.W. Littlejohn	1953-71 Godfrey Phillips
C.F.B. Paul	1953-64 St James Tobacco Co.
J.W. Turner	1953-54 National Tobacco Co.
P.J. Taljaard	1958-60 Rothmans New Zealand
C.C. Chater	1959-63 W.D. & W.O. Wills
L.P. Hamlin	1960-70 Rothmans New Zealand
P.G. Fulton	1963-75 W.D. & W.O. Wills
A.J. Mitchell	1964-66 and 1975-85 W.D. & W.O. Wills

J.G. Husheer   1966-80 Rothmans New Zealand
K.D. Butland   1970-72 Rothmans New Zealand
M.W.M. Rouse   1971-76 Phillip Morris
G.R. Fraser   1972-74 Rothmans New Zealand
E. Bradley   1976-80 Phillip Morris
L.F. Doolan   1980-86 Rothmans New Zealand
P.U. Lovatt   1981 Phillip Morris
D.J. Weaver   1986 W.D. & W.O. Wills
G.B. Drummond   1986 Rothmans New Zealand

Secretaries

H.L. Wise   1936-69 (secretary and chairman 1958-69)
Sylvia M. Chesney   1968-77 (assistant secretary 1959-68, acting secretary 1978)
L. Coleen Tomlins   1977-78
J.A. Harding   1978 (February-June)
E.H. Clayton   1978-82
J.J. (Jill) Moss   1982-86

Government representatives

L.J. Schmitt   1936-53   Department of Internal Affairs
H.L. Wise   1936-69   Department of Industries and Commerce
E.J. Fawcett   1938-53   Department of Agriculture
K. Laurence   1947-49   Rehabilitation Department
W. Andrew   1947   Rehabilitation Department
E.M. Basil-Jones   1948-55   Rehabilitation Department
J.H. Whiteford   1955-56   Rehabilitation Department
G. Laurence   1965-70   Department of Industries and Commerce
J.W.H. Clark   1970-73   Department of Industries and Commerce
P.J. McKone   1973-75   Department of Trade and Industry
M.G. Baumgart   1974-86   Ministry of Agriculture and Fisheries
D.J. Gasson   1975-83   Department of Trade and Industry
D.L. Shroff   1983-86   Department of Trade and Industry
B.G. Koller   1986   Ministry of Agriculture and Fisheries

Tobacco Production in Motueka 1925-95

Year	No. of Growers	Acreage	Production
1925-26	160	450	No figures available
1926-27	140	300	"
1927-28	314	720	"
1928-29	632	1608	(Includes 73 growers in other areas)
1929-30	506	1587	(Includes 10 growers and 5 companies in other areas)
1930-31	Not known	932	722,329 pounds (56,077 outside Nelson)
1931-32	"	1726	1,318,624 pounds (93,286 outside Nelson)
1932-33	"	2126	1,784,676 pounds (239,677 outside Nelson)
1933-34	"	1803	1,239,946 pounds (204,320 outside Nelson)
1934-35	"	1358	1,106,424 pounds (100,000 outside Nelson)
1935-36	499	1969	1,197,161
1936-37	508	2770	1,609,493
1937-38	442	2563	2,067,827
1938-39	342	2225	1,423,610
1939-40	339	2570	2,217,040
1940-41	381	2963	3,143,355
1941-42	381	2875	2,721,539
1942-43	360	2907	3,185,183

1943-44	414	3066	3,083,094
1944-45	487	3303	3,286,067
1945-46	553	3405	4,080,135
1946-47	628	3805	4,706,723
1947-48	661	4322	4,770,827
1948-49	619	4393	5,000,470
1949-50	541	3899	4,711,613
1950-51	529	3904	5.436,314
1951-52	496	3648	4,088,189
1952-53	429	3514	4,797,267
1953-54	401	3212	4,228,643
1954-55	412	3082	4,135,921
1955-56	418	3137	4,733,086
1956-57	416	3138	4,463,976
1957-58	411	3265	4,650,711
1958-59	437	3534	5,606,951
1959-60	468	3750	7,075,577
1960-61	549	4151	6,777,400
1961-62	613	4699	9,327,381
1962-63	729	5357	8,947,789
1963-64	763	5878	9,380,752
1964-65	728	5840	9,880,814
1965-66	585	5088	6,822,551
1966-67	529	4882	5,515,631
1967-68	526	5101	6,154,140
1968-69	523	5067	7,603,906
1969-70	506	5023	7,161,579
1970-71	464	4934	8,571,823
1971-72	411	4636	7,333,808
1972-73	353	4296	6,712,057
1973-74	340	1751ha 4377 acres	3,237,779kg (7,123,113lbs)
1974-75	306	1764	3,006,772
1975-76	308	1775	3,349,115
1976-77	315	1847	2,718,551
1977-78	309	1895	3,558,965
1978-79	301	1718	3,749,820
1979-80	269	1513	3,250,544
1980-81	237	1027	2,290,293
1981-82	122	470	1,506,362
1982-83	135	657	1,583,085
1983-84	143	735	1,674,710
1984-85	131	814	2,009,902
1985-86		824	1,775,595
1986-87	119	734	2,027,127
1987-88	100	544	1,217,930
1988-89	91	519	1,503,813
1989-90	57	339.8	896,400
1990-91	53	308.36	891,539
1991-92	53	334.36	943,810
1992-93	53	313.61	838,279
1993-94	51	217.95	655,593
1994-95	51		

Figures to 1934-35 taken from New Zealand Yearbooks and information supplied to the 1930 Committee of Inquiry. Figures from 1936 taken from Tobacco Board annual reports – these do not always agree with New Zealand Yearbook entries.

Average Price of Flue-cured Tobacco Leaf

1936-37	1s 7 1/2d	1956-57	4s 1 1/2d	1976-77	$3.33c
1937-38	1s 8d	1957-58	4s 1 1/2d	1977-78	$3.46c
1938-39	1s 10 1/2d	1959-60	4s 4d	1978-79	$3.85c
1939-40	1s 10 1/2d	1960-61	4s 4d	1979-80	$4.20c
1940-41	1s 10 1/2d	1961-62	4s 5d	1980-81	$4.45c
1941-42	1s 10 1/2d	1962-63	4s 5d	1981-82	$4.80c
1942-43	1s 10 1/2d	1963-64	4s 5d	1982-83	$5.25c
1943-44	1s 11 1/2d	1964-65	4s 7 1/2d	1983-84	$5.44c
1944-45	2s 0 1/2d	1965-66	4s 9 1/2d	1984-85	$5.80c
1945-46	2s 2d	1966-67	5s 5d		(approx 26s 4d lb)
1946-47	2s 2d	1967-68	58c (5s 9d)	1985-86	$6.00
1947-48	2s 5 1/2d	1968-69	66c	1986-87	$6.14
1948-49	2s 5 1/2d	1969-70	66c	1987-88	$6.70
1949-50	2s 8d	1970-71	72c	1988-89	$6.56
1950-51	3s 0d	1971-72	78.5c	1989-90	$6.25
1951-52	3s 2 3/4d	1972-73	88.5c	1990-91	$6.50
1952-53	3s 6 1/2d	1973-74	$1.01c	1991-92	$6.60
1953-54	3s 10 1/2d	1974-75	$2.50c per kilogram	1992-93	$6.65
1954-55	4s 0d		($1.14c per pound)	1993-94	$6.65
1955-56	4s 1d	1975-76	$2.80c	1994-95	$6.65

Sources of Reference

PUBLISHED

Beatson, C.B., *The River, Stump and Raspberry Garden*, Nikau Press, Nelson, 1992.
Beatson, K. and Whelan, H., *The River Flows On*, Beatson and Whelan, Nelson, 1993.
Belich, J., *I Shall Not Die*, Allen and Unwin, Wellington, 1989.
Brereton, C.P., *Vanguard of the South*, A.H. & A.W. Reed, Wellington, 1952.
Chippendale, F. et al, *Tobacco Growing in Queensland*, Queensland Department of Agriculture and Stock, 1961.
Dovedale's History, compiled by Dovedale Agriculture and Craft Centre Committee, 1990. Supplied by Olga Barker, Stanley Brook.
Dunlop, B. and Mooney, K., *Profile of a Province – Hawke's Bay*, Hodder and Stoughton, 1986.
Eldred-Grigg, S., *Pleasures of the Flesh*, A.H. & A.W. Reed, Wellington, 1984.
Mack, Peter H., *The Golden Weed*, Millbrook Press, Southampton, 1965.
Newport, J.N.W., *Footprints Too*, Express Printing Works, Blenheim, 1978.
Promotional brochure produced by the Empire Tobacco Company, c. 1930.
Sutherland, G.F., *Tobacco Growing in New Zealand*, Government Printer, 1901.
The Culture of the Tobacco Plant, translated by order of Governor Sir George Grey, Auckland, 1867.
Thomson, R., *Flue-cured Tobacco Growing in New Zealand*, Department of Scientific and Industrial Research, 1948.
Tobacco Australia, Australian tobacco industry publication, c. 1970.

NEWSPAPERS AND PERIODICALS

Daily Telegraph, Napier
Marketplace, 1972
Motueka-Golden Bay News
Motueka Star-Times
Nelson Evening Mail
New Zealand Farmer Stock and Station Journal, 1913, 1914, 1921
New Zealand Free Lance, April 1928
New Zealand Tobacco Growers' Journal
New Zealand Truth
Straight Furrow, August 1980
The Dominion, Wellington
Weekend Star

UNPUBLISHED

A.L. Black, draft of evidence prepared for Wills case, c. 1990.
Cattermole, G., W.D. & W.O. Wills (NZ) Ltd Motueka Processing Plant (1942-1989), school project, Motueka High School, 1992.
Heath, E., Reminiscences of Life in Tobacco, Especially Stories of Workers.
Holyoake, H., Notes on the Situation of the Industry, c. 1954.
O'Shea, P.K., The Impact of Deregulation on the Tobacco Growing Industry, second-year sociology paper, University of Canterbury, 1993.
Rowling, G., Biographical notes, supplied by Sally Goodall, Riwaka.
Tillson, G., Notes on the History of Tobacco in NZ, 1980.
Waugh, J.R., The Changing Distribution of Tobacco Growing in Waimea County, MA thesis in geography, 1962.

OFFICIAL DOCUMENTS

Appendices to the Journals of the House of Representatives (AJHR).
Copies of submissions to select committee on Tobacco Growing Industry Bill, 1973.
File of the proceedings of the Committee of Inquiry into Tobacco Growing, 1930. National Archives file No. 1930/16.
New Zealand House of Representatives and Legislative Council Parliamentary Debates.
New Zealand Statistics.
New Zealand Yearbooks 1885-1993.
Records of the Tobacco Board, National Archives.
Report of the Committee of Inquiry into Tobacco Growing, 1930.
Report of inquiry in Tobacco Manufacturers, 1936, National Archives.
Report of the Inter-departmental Study Group, 1980.
Report of Select Committee on Industries 1919, National Archives.
The Economic Position of the Tobacco Growing Industry in New Zealand, report of the Committee of Inquiry, Government Printer, 1971.
Tobacco Board Annual Reports 1936-1986.
Transcript of the Committee of Inquiry into the Tobacco Growing Industry 1966-68. Supplied by J. Silcock, Motueka.
Various copies of Tobacco Board minutes.

OTHER SOURCES

Archives of Rothmans of Pall Mall (New Zealand) Ltd, Price Waterhouse Centre, Auckland.
Copies of television interviews, both broadcast and unbroadcast. Execam, Wellington.
Interviews with growers and former growers throughout the Motueka area.
Interviews with descendants of Pongakawa settlers.
Interviews with former company personnel.
Interviews with research personnel.
Interview with Eric Highet, Tobacco Board chairman.
Interviews with grower members of the board: Maurice Cederman, John Hurley and Gerald Hunt.
Interviews with members of local community, including police and doctors.
New Zealand Tobacco Growers' Federation records, including minutes, annual reports and correspondence plus selected papers of R.W.S. Stevens, scrapbooks, ward minutes and miscellaneous material.
New Zealand Tobacco Growers' Cooperative Society records, including minutes, annual reports and correspondence.
Notes on New Zealand Workers' Union, supplied by John Krammer, Motueka.
Papers supplied by John Husheer, Napier, including:
    Copies of articles on the history of Riverhead;
    New Zealand Tobacco Company Ltd Company Prospectus
    Notes on the History of the Tobacco Industry, compiled by T. Husheer;
    'New Zealand Tobacco Farm', publisher and date unknown;
    Correspondence between Cecil Nash and Gerhard Husheer.
Personal papers of Charles Lowe, made available by R. Lowe, Wellington.
Personal papers supplied by G. and K. Emerre, Riwaka.
Personal papers of Fred A. Hamilton, supplied by Coralie Smith, Motueka
Personal papers of Rona Hurley and F.O. Hamilton, supplied by J.C. Hurley, Pangatotara.
Personal papers supplied by K. and B. Mytton, Motueka.
Photocopies of articles on Pongakawa supplied by June Drabble, Pongakawa including Potter, Phyllis 'History of Rotoehu School'.
Telephone discussion with Harry Cardiff, 2nd Field Regiment, Second World War.
W.D. & W.O. Wills scrapbooks and correspondence held in Motueka Museum.

Index

Adams-Schneider, L., MP 176, 177, 178, 180, 182, 184, 185
Addison, J. 18
Africa 29
Agricultral Workers' Union 134
Agricultural Training Council 191
Allen, H.J. 141
Allen, J.M. 80, 104
American Tobacco Company 11, 12
Ankerschmidt & Company 14
Arataki 14
Arawa Trust 38, 39, 58
Arbitration Court 93, 102, 119, 141
Armstrong, H.T., MP 50, 80, 83
Ashby, S.F. 174
Askew (growers) 27
Atkins (growers) 27
Atkinson, G. 18
Atmore, H.A., MP 51, 68
Auckland 14, 15, 37, 38, 39
Auckland Tobacco Company 13, 17
Australia 20, 27, 29, 39, 49, 52, 57, 104, 196
Australian Department of Agriculture 80
Awakeri 138
Awanui 14
Awhatu 38
Baas, Colin 155
Balck Bridge 63
Balck, James 45, 51, 62, 63, 71, 72, 75, 86,
Barham, Justin 80
Barker, Olga 92, 107
Barker, Tom 107
Barlett, A.E. 41
Bartlett, Elizabeth 186
Barton, George, QC 202, 204
Bassett, Dr E. 119, 125
Bate, Ralph 23
Bates, Percy 23
Baumgart, Moore 190
Bay of Islands 40
Bay of Plenty 12, 13, 37, 38, 43, 53, 57, 58, 138, 142
Beatson, George 52
Beatson, Pat 35, 66
Beatson, R. 162, 171
Beckett, Darien 204
Bedi, A.S. 162
Bell, A.F. 76
Bell, Captain Allen, MP 37
Benner, Renee 58
Bennetts Fred 65
Bensemann (growers) 27
Bensemann, C.M. 46
Best (growers) 27
Best, Alex 46
Beuke (growers) 27
Black, A.L. 22, 104, 108, 147, 148, 161, 196, 197, 203

Black, George, MP 43, 46, 48, 49, 50
Blenheim 41
Bloomfield's 66
Bloomfield, E.H. 113
Bolger, Len 196
Bond & Bond Ltd 15
Bond, Enos 15
Bond, W. 159, 175
Bonsack, James 10
Boyden, W.E. 103
Bradley, E. 104, 161
Bradley, J.J. 116, 130, 144
Brame, Sedrick 26, 37, 45, 46, 104
Branford (nurseries) 30
Brereton, Cyprian 23, 26, 51, 71, 74, 76, 86
Brereton, William 17
Brightwater 20
Brislane, W. 63
Brislanes (growers) 27
Britain 10, 11, 49, 69, 79, 98, 178
British American Tobacco Company (BAT) 11, 15, 19, 48,
British Empire Trading Company 20
Brodie (instructor) 26
Brodie, J.W. 15, 18
Bromley, F.J. 79, 80
Brooklyn hydro-scheme 64
Brough (nurseries) 30
Brown, Mark 207
Bruce, J.L. 113
Bryson, Frank 150
Buck, E. 12
Burley 29, 95
Burnett, Fen 65
Burnett, S.E. 46
Burnett, W.H. 27, 36
Burns, John 15
Butcher, D., MP 196
Butland, K.D. 166, 167, 169
Buxton, E. & Company 20, 24, 45, 48, 52, 73, 81, 83
Campbell, J.A. 24, 25, 40, 43, 44, 57, 79, 210
Canada 11, 26, 104
Canterbury Agricultural College (Lincoln) 104, 107
Canton, John 52
Canton, T.D. 159
Cape Province 12
Carruthers, Colin 200, 201
'Cavendish Mixture' 15
Cawthron Institute 80, 162
Caygill, D., MP 190, 195, 211
Cederman, David 186
Cederman, Maurice 143, 157
Cederman, Russell 199
Central Committee of Tobacco Manufacturers 151, 160
Charles I 10

Chesney, Sylvia 97
China 9, 29
Clark, P. 114
Closer Economic Relations Agreement (CER) 191, 202
Clouston, Ian 179
Coates, G., MP, Prime Minister 52, 63, 69, 70, 71
Collins, K.C. 144
Columbus, Christopher 9
Commerce Act 191
Commerce Commission 191
Commission of Trade Prices and Practices 128
Commission on Primary Industries 78
Committee of Inquiry into the Tobacco Growing Industry in NZ 1966-68 149, 160, 162, 163, 166, 167, 211
Committee of Inquiry into Tobacco Growing 1930 43, 46, 53, 70
Commonwealth Scientific and Industrial Research Committee 80
Consolidated Tobacco Company 75, 83
Cook, Fred 23
Cooper, Warren, MP 188
Corder, J. 67
Country Women's Institute 65
Cowan, James 14
Cox, M.D. 191
Craig, Dr 53
Crimean War 10
Crisp, Dennis 177, 179
Crop Insurance Reserve Fund 96
Curnow (grower) 72
Customs Tariff Commission 62, 63
'Cut Plug No 10' 15
Daken, E. 142
Dansey, Rodger 38, 39
Davies (growers) 27
Davy, V. 86
Deck's Reserve 83, 84
Delany, Hamilton 18, 22
Department of Agriculture 14, 17, 24, 25, 27, 37, 40-2, 49, 53, 55, 57, 80, 82, 83, 84, 110, 112, 113, 120, 122, 128, 163
Department of Agriculture Act 1953 140
Department of Education 56
Department of Industries and Commerce 49, 75, 80, 82, 103, 112, 114, 128, 133
Department of Labour 53, 55, 83, 101, 116, 117, 119, 141
Department of Maori Affairs 117
Department of Scientific and Industrial Research 80, 108, 171, 193
Department of Trade and Industry 172, 175
Dickie, Chris 198, 207
Disabled Soldiers Civil Re-establishment League 79
Discharged Soldiers Act 65
Dockery, R.C. 137
Doidge, F.W., MP 99
Dominion Tobacco Company 13, 19
Donaven, Dennis 14
Doolan, Laurie 185, 189

Douglas, George 36
Dovedale 25, 27, 31, 33, 36, 45, 46, 47, 50, 60, 62, 63, 65, 74, 75, 78, 99, 119, 132, 163, 192
Drabble, June 58, 59
Drummond, E.W. 183, 191, 192, 194, 197, 199-202, 208
Drummond, J.R. 113, 115, 119
Drummond, L.K. 46
Drury and Bradley (accountants) 116
Dublin 26
Duff, Allan 142
Duke, W, Sons & Company 10
Duncan, John 45, 51
Durham 10
Dutton, Horace 129
Ebbet, George 14
Eden, Gilbert 87
Eggers, Dudley 27, 66, 107
Eginton (growers) 27
Eginton, Walter 114, 115, 123, 142, 143, 145
Emerre, Graeme 144, 204
Emerre, S.J. 131
Empire Tobacco Corporation 39, 43, 44, 58, 59
Europe 9
Evans, W. 62
Everett, Boyd 176
Everett, Horatio 17, 52, 84, 88
Everett, Hubert 68
Farmers' Union 68
Federated Farmers 122
Federation of Labour 179, 187
Feilding 103
Feist, R.T. 202, 203
Fijians 170
Forbes, G., MP, Prime Minister 52, 53
Forestry Department 56
Forsyth, A. 18
Fry (growers) 27
Fry, A.C. 200
Fry, Alfred 61-63, 70-72
Fry, B.Y. 66, 200
Fry, Cyril 23, 51
Fry, K.J. 186
Fry, Melvyn 132
Fry, Stanley 47
Fry, W. 157
Gardiner, E.J. 125
General Tobacco Company 75
Gibbons (farmer) 42
Gibbs, F.W. 23, 26, 27, 48, 54
Gilmour (instructor-director) 15, 20, 26
Gisborne 142
Gloag, Robert Peacock 10
Glynan, M.M. 141, 142
Godfrey Phillips 81, 104, 134, 136, 137, 138, 160
'Gold Pouch' 15
Good, E.W. 48
Goodall, Russ 107, 131, 135
Goodall, Sally 35
Gotch, W 12
Green, G.A. 43
Grey, Sir George 12
Growers' Federation 88, 106, 129

Haldane, J.R. 99
Hall, Lance 65
Hall, Ruth 65
Halstead, E.H., MP 128, 129
Hambleton, Betty 28
Hamilton, A., MP 64
Hamilton, Fred A. 51, 71, 74, 75, 86, 87, 89, 91, 96, 98, 101, 106, 111-15, 119, 131, 140
Hamilton, Fred O. 14, 20, 22, 24, 48, 51, 52, 65, 68, 73, 74, 81, 82, 84
Hamilton, Ian 35, 64, 104
Hamilton, J.P.B. 157
Hamlin, L.P. 167, 168
Harakeke 17, 18, 19, 21
Hartlaub & Company 14
Hartshorn, Federick 14
Harvey, C.L. 45, 68
Haumoana 14, 15
Havana 9
Havelock North 14
Hawke's Bay 12-15, 30, 142
Hawkes (growers) 27
Hawkes, W.M. 45, 54
Heaps, Cyril 63, 96
Heath, Beth 118
Heath, Cyril 45
Heath, E. 125
Heath, J.T. 45
Heath, Mervyn 179, 181
Heath, Murray 118
Heath, Peter 144, 148
Heine, Pastor 17
Helensville 39
Helm, H.M. 25
Hickmott (growers) 27
Highet, E.S., MP 178, 179, 181, 186, 187, 190
Hill, F. 20
Hobson 13
Hodgkinson (growers) 27
Hodgkinson, R. 45
Hokianga 37, 38, 41
Holland, S.G., MP, Prime Minister 107
Holloway, N.C.L. 144
Holloway, P.N., MP 130, 133
Holyoake, F.T. 101
Holyoake, Harold 120, 124, 127, 129, 134,
Holyoake, K.J., MP, Prime Minister 50, 51, 53, 61, 63, 64, 67, 69, 70, 73, 74, 78, 79, 84, 86, 87, 88, 110, 114, 122, 145, 150, 169, 172, 210, 211
Hopwood, B. 93
Hortresearch 17
Hunt, E.M. 76, 96
Hunt, Gerald 143, 144, 157, 165, 169, 180, 184, 186, 187, 189, 191, 200, 202, 204, 205, 211
Hurley (growers) 27
Hurley, Daniel 46, 145
Hurley, John 135, 144, 145, 147, 148, 157, 165, 167, 175, 180, 182, 184, 186, 188, 200, 206
Hurley, Rona 76, 81, 90, 97, 103, 142, 145, 147
Husheer, Gerhard 14, 15, 18, 20, 21, 22, 24, 30, 37, 39, 47, 70, 71, 73, 75, 76, 84, 108, 136, 142, 209

Husheer, Ingolf 136, 137
Husheer, John 15, 147, 180
Husheer, Torvald 16, 18, 43, 136, 147
Imperial Tobacco Company Ltd 10, 19
Industrial Conciliation and Arbitration Act 134
Industries Development Commission 185-7
Inglis (growers) 27
Inglis, Jack 175, 193
Inglis, Tom 177, 183, 194, 196
Inwood (growers) 27
J.E. Trevett New Zealand Ltd 183
James (growers) 27
James, R.W. 139, 153, 154, 162, 170, 171, 181, 190
Japan 26
Java 9
Jaycees 144
Jenkins (growers) 27
Jenkins B.W. 45
Jones, William 14
Jordan, C.E. 45, 87
Jordans (growers) 27
Jury, A.A. 104
Kaikokopu 57
Karaitiana, Waikare 14
Kawhia 13
Kempthorne Prosser 31
Kennedy, David 182, 187
Kenyon, A. 67
Kenyon, W.G. 113, 129, 130, 139
Kerr, Roger 186
Kerr, W.G. 87
King James I 10
Kirk, Alf 187
Kirk, T.W. 17, 18
Kohu Kohu 38
Kouto 38
Krammer, John 177-79
Kumeu 39
Kyle, T.A. 142
Lands and Survey Department 65
Laurence, George 147, 149, 150, 151, 154, 155, 163, 167-69, 172
Legg, Robert 10
Leppien, L.W. 114, 115
Lewis, Griffith 20
Lewis, Noel 68-71, 87, 93, 100, 111, 113, 128, 130, 150, 151, 153, 154, 155
Lindeman, Bruce 137
Littlejohn, F. 134
Littlejohn, P.B. 104, 160
Lloyds (insurance) 96
Lock, C.P. 43, 48
London 17, 38, 81, 84, 102
Lough (instructor) 26
Lowe, Charles 17-20, 22, 24, 27, 32, 33, 35, 37-43, 46, 53, 55-58, 79-81, 83, 84, 209, 210
Lowe, R. 35
Lower Moutere 27, 45, 60, 88, 117
Loyals Tobacco Company 51, 57
Lusty, Bernard 46
MacMillan, D.G., MP 73
Malawi 11, 146

Manoy and Sons 23
Manoy, Harry 61, 70, 75
Manukau 13
Maori 12, 15, 37, 38, 39, 117
Mapua 18, 27
Marahau 27, 65, 66, 98
Marlborough 38, 41, 46, 51
Marshall, J.R., MP, Prime Minister 140, 145, 157
Marshall, Kerry 205
Marshall, P.B. 160, 166
Martin, Jack 101, 113
Martin, John 42
Martin, Pat 36, 44, 108, 124
Martin, W.L., MP 78
Masters, R., MP 71, 73
Masterton 138
Maude, Dorothy 56
Maude, Percy 55
Maxwell, James 51, 60
McGee (Nelson) 17
McGlashen (growers) 27
McGlashen, J.A. 47
McGlashen, J.F. 63
McGregor, A.G. 142
McKay, Lindsay 171
McLaren, H.A. 128, 140
McLean, A.J. 203
Mexico 9
Milnes, L.A. 154
Minor Company Growers' Committee 85
Moffat, W.J. 67
Molloy, A, QC 198, 201, 203, 204, 207
Moore, G. 192
Morea 38
Morgan, K.A. 125
Moss, Alan 208
Moss, Frank 23, 26
Motueka Beach Camp 92
Motueka Borough Council 123
Motueka District Tobacco Growers' Association 37, 55, 60, 114
Motueka Harbour Board 68
Motueka Progress League 68
Motueka Public Library 153
Motueka River 64, 88, 105
Motueka Valley 23, 24, 109, 118, 163, 192, 196, 212
Motueka, passim
Moutere 50, 51
Moutere Valley 163
Muldoon, R.D., MP, Prime Minister 182, 183
'My Favourite' 15
Mytton, Kelvin 66, 125
Mytton, Lance 23, 66
Mytton, Mona 110
Napier 14, 15, 20, 27, 42, 46, 103, 137
Nash, Cecil 20-25, 27, 29, 31, 35, 34, 39, 43, 44, 46, 47, 60, 61, 72, 74, 76, 78, 81, 102, 104, 108, 209
Nash, W., MP, Prime Minister 97
National Services 92
National Tobacco Company 15, 16, 18, 20, 23-27, 30, 32, 38, 41, 42, 46-49, 51, 61-64, 70, 74, 75, 81, 82, 84, 85, 86, 89, 102, 103, 136, 145, 209
'Navy Cut No 3' 15
Nayasaland (Malawi) 11
Nelson City Council 68, 122
Nelson Evening Mail 93
Nelson Pioneer Tobacco Growers' Union 85
Nelson Provincial Progress League 68, 123
Nelson Provincial Tobacco Growers' Association 71-75, 85
Nelson Tobacco Company 76, 102, 104
Nelson Trades Council 187
Nelson, passim
Nelson-Bays United Council 187
New Brighton 79, 80
New Zealand Tobacco Company 14, 15, 18, 20, 39, 40
New Zealand Tobacco Growers' Cooperative Society 194
New Zealand Tobacco Growers' Federation 40, 85
New Zealand Tobacco Growing & Curing Company 13
New Zealand Workers' Union 88, 93, 119, 141, 143, 177, 178, 183, 187
Newman, K.J. 112, 115, 127, 129, 131, 143, 148
Ngatimoti 23, 52, 112
Nordmeyer, A. MP 99
North Carolina 10
North Island 41, 51, 61, 84
Northland 37, 55
O'Flynn, F.D., MP 89, 150, 151, 155, 160, 171
O'Hara, Sam 65, 110, 120, 133
Ohinemutu 15
Omanaia 38
Opotiki 13
Orinoco 23-26, 35, 51, 60, 109, 163
Ormond, J.D. 13
Oronoko 18
Otaki 103
Ottawa Imperial Conference 1933 52, 53
Owen, E.V. 51
Owen, W.H. 23, 42, 43, 44, 57
Pacific Tobacco Company 40
Pakipaki 14
Palmer (grower) 72
Palmerston North 80
Pangatotara 25, 27, 51, 60, 61, 62
Papakura 12
Papamoa 39
Park, D.R. 62, 68
Parkes (growers) 27
Parkes, Joshua 23
Paroa 138
Pearse River 23
Perry, George 46
Pethybridge, C.C. (Charles) 103, 160
Petone 34, 38, 39, 46
Phillip Morris 160, 169, 179, 180, 188, 196
Phillips, Arthur 81
Pigeon Valley 65
Pikowai 57
Pioneer Union of Tobacco Growers 75

Plunket Society 65
Pokororo 17, 25, 45, 60
Pokororo Hall 92
Pongakawa 58, 80, 81, 84
Pongakawa Valley 42, 53-58, 210
Pope, D.E. 93
Port Ahuriri 14, 23, 32
Post Office Saving Bank 119
Price Stabilisation Tribunal 110
Price Tribunal 99, 100, 110, 120, 122, 123, 127, 128, 130, 129, 133, 136, 147, 151, 152, 156, 157, 159, 160, 171
Primmer (growers) 27
Public Works Department 94
Puhoi 14
Queen Elizabeth I 10
Queensland 11, 52
Ramsay (sales rep.) 64
Rance, Jeanne 175, 191
Rance, Roy 175, 181, 191
Rangitikei 13
Ravensdown 31
Reakes, Dr C.J. 43
'Red Shag' 15
Rehabilitation Scheme 143
Reid, John 176
Relat, George 75, 79, 86
Rembrandt of South Africa 136
Rhodesia (Zimbabwe) 11, 20, 146
Richardson, R., MP 177
Richmond 46
Richmond Borough Council 68
Richmond Showgrounds 92
Riddiford, D.J., MP 168
'Riverhead Gold' 15
Riverhead 15, 22, 24, 30, 39, 40, 46
Riwai, Kia 117, 118, 142
Riwaka 23, 24, 26, 34, 39, 40, 45, 46, 47, 60, 63, 64, 65, 78, 80, 88, 94, 98, 99, 102, 113, 131, 132, 142, 158, 163, 177, 186, 193
Riwaka River 105
Ross, Ian 194
Rothmans Tobacco Company Ltd 15, 103, 136-38, 142, 145, 146, 152, 166, 167, 169, 172-74, 176, 177, 184, 185, 186, 189, 195, 199, 201, 204-07, 210, 211
Rotoehu 56
Rotorua 12, 15, 38, 39, 55, 58
Rouse, M.W.M. 169
Rowling (growers) 27
Rowling, B.T. 36, 51, 61, 62, 70, 75, 88
Rowling, Gordon 34, 35, 40
Rowling, Nolan 75, 84, 85, 96, 113, 114
Rowling, Syd 24, 52
Rowling, W.E., MP, Prime Minister 86, 87, 157, 159, 160, 163, 166, 168, 171-73, 178, 183-87, 190, 211
Ruatoki 138
Ruby Bay 18, 36
Rural Bank 172
Russia 9, 26
Rutherford Farm 20
Rutledge and March (growers) 51

Sandy Bay 27, 41
Savage, M.J., MP, Prime Minister 74, 87
Schmitt, Louis 75, 86, 101, 107, 111, 113, 114
Selby-Bennetts, R.F. 62, 68
Select Committee on Industries 1919 18
Semple, R., MP 74, 75
Shelly Beach 39
Shelton, N., MP 167, 168
Shirley, K., MP 192, 196
Shuttleworth, S. 72
Silcock C.A. 27
Skillicorn, A. 157
Skinner, C.F., MP 97, 112, 130, 133
Skipper, W.H. 125
Small Farms Board 65
Small Farms Scheme 53, 55, 63
Smallover, S. 15
Smith, Coralie 108
Smith, R.B. 36, 47, 48, 54, 58, 59, 75, 76, 118
Smith, S., MP 130
Smith, Spencer 22, 51, 53, 60, 61, 62, 63, 64, 67, 69, 70, 71
'Smoke-ho' 15
Smythe, Dr 37
Snedden, K.A. 75, 76
Sorenson, Richard 14
South Africa 12, 146
South America 9
South Carolina 17
Spain 9
Spanish Armada 10
Spratt, A.H. 75
St. James Tobacco Company 102, 103, 144
St. Louis 17
St. Louis Exhibition 17
Standard Tobacco Company 39
Stanley Brook 23, 26, 60, 68, 92, 107, 109, 163
Staples (growers) 27
State Advances Department 65
Stephens (growers) 27
Stevens, R.W.S. (Dick) 86, 89, 99-103, 111, 114, 147, 149-3, 155-60, 166, 167, 169, 173, 175
Stevens, John 36, 108, 125
Stott, (sales rep.) 64
Strachan, Cara 135
Stringer, H.G. 75, 79
Sullivan, D.G., MP 75
Sunrise Valley 27
Sutherland, G.F. 14
Sutherland, Larry, MP 187
Tadmor 78
Tait, Peter 137
Takaka 41
Talbot, Rowly 142, 143, 145, 156
Taljaard, P.J. 137
Tanner, Thomas 12
Tapawera 23, 26, 48, 60, 65, 72, 98, 109, 163
Taranaki 12, 13, 15
Tariff Commission 53, 67, 69, 70, 71
Taumarunui 103
Tauranga 38, 57
Tauranga Citrus and Tobacco Company 39, 43, 48, 57

Taylor, A.D. 68, 84
Taylor, Eleanor 69
Te Atatu 15, 37, 38, 39, 46, 49
Te Puke 53, 55
Te Teko 138
Tengi 38
Tennent, R.B. 117, 122, 123, 127, 133, 137
Thomas, Fred 132
Thomas, Godfrey 113
Thomason, H.H. 142
Thomson, Robert (Bert) 104, 162
Thorn, Hugh 27, 75
Thorn, Martin 110
Thorn, Maurice 87
'Three Diamonds' 15
Tiffen, H.S. 12
Tillson, Geoff 206
Tobacco Act 1908 49
Tobacco Advisory Council 49, 50, 128, 140
Tobacco Board 73, 75, 78-88, 91, 97, 98, 100, 105-7, 109-12, 114-16, 120, 122, 126-30, 133, 136, 139, 141, 143-47, 150, 151, 158-60, 163, 166-69, 172, 173, 175-79, 184, 188, 190, 193
Tobacco Companies Trust 142
Tobacco Growers (NZ) Ltd 40
Tobacco Growers' Asoscciation 51, 62, 71, 86, 96
Tobacco Growers' Cooperative Society 202
Tobacco Growers' Federation 60, 110
Tobacco Growing Centre 38
Tobacco Growing Industry Act 1935 74, 81, 85, 127
Tobacco Industry Amendment Act, 1981 188
Tobacco Industry Bill 73
Tobacco Institute 189
Tobacco Producers' Trust 39
Tobacco Quota Committee 172
Tobacco Research Station 104, 131, 153, 162
Tom Thumbs, 10
Transport Nelson 106
Umukuri 51, 52, 62, 63, 64, 104
Unemployment Board 53, 65
United Tobacco Corporation (Tauranga) Ltd 39, 57
Upper Moutere 17, 27, 45, 60
Upper Takaka 27
USA 11, 19, 26, 29, 104, 178
Victoria 11
Virginia 9, 10, 17
Virginia Gold 105, 132, 140, 154
Vogel, Sir Julius 12
Volbracht, A. 13
Wai-iti 26, 109, 163
Waikanae 17
Waimea 38, 45, 60, 72, 132, 138, 163

Waimea County 13
Waimea County Council 67
Waimea Electric Power Board 68, 123
Waipa 13
Wakapuaka Cemetery, Nelson 21
Wakefield 65, 119
Walker, Ethena A. 162
Walsh, Austin 13
War of Independence 10
Watt, P. 18
Watterton, Stuart 189
Watts, J.T., MP 113
Wearne, Trevor 183, 184
Weaver, David 189
Webb, P.C., MP 68, 83
Wellington 62, 73, 74, 103
Wellsford 40
Werthmuller, Eugen 176
West, H.G. 154
Western Bay of Plenty 55
Whakarewarewa 11
Whakatane 138
Wheeler, Mark 200, 206
Whirinaki 38
White Burley 18
Whitehead, S., MP 168
Whittaker (instructor) 26, 37
Whittaker, T.J. 19
Wild, Peter 104, 144, 157, 161
Wilkinson, A.C.M. 18
Wilkinson, R.E. 181
Williams, Bob 107
Williams, Hugh 63
Williams, R.J. 36
Wills, W.C. (Fred) 107, 110-15, 126, 128-30, 133, 134, 139, 140, 142
Wills, W.D. & H.O. (Bristol) 10, 15
Wills, W.D. & H.O. (NZ) Ltd 18-20, 24-27, 32-38, 41, 42, 43, 45-51, 57, 58, 64, 70, 72-74, 82, 84, 85, 102-4, 136, 138, 142-47, 149, 173, 174, 176, 179, 189, 190, 194-6, 198, 201, 203, 204, 206, 207, 209, 210
Wilson, Cecil 123
Win, W.S. and Son 45
Wins (growers) 27
Wise, Henry 97, 100, 101, 107, 113, 128, 130, 133, 134, 138, 139, 145, 146, 152, 157, 160
Women's Division of Federated Farmers 65, 68
Woodstock 27, 51, 60
Yellow Pryor 18
York, Rupert 52, 67, 68
Young Farmers' Club 144
Zambia 162
Zimbabwe 11